POWER SYSTEM
TRANSIENT ANALYSIS

POWER SYSTEM TRANSIENT ANALYSIS
THEORY AND PRACTICE USING SIMULATION PROGRAMS (ATP-EMTP)

Eiichi Haginomori
University of Tokyo, Japan

Tadashi Koshiduka
Tokyo Denki University, Japan

Junichi Arai
Kougakuin University, Japan

Hisatochi Ikeda
University of Tokyo, Japan

This edition first published 2016
© 2016 John Wiley & Sons, Ltd.

Registered Office
John Wiley & Sons, Ltd, The Atrium, Southern Gate, Chichester, West Sussex, PO19 8SQ, United Kingdom

For details of our global editorial offices, for customer services and for information about how to apply for permission to reuse the copyright material in this book please see our website at www.wiley.com.

Library of Congress Cataloging-in-Publication Data

Haginomori, Eiichi, author.
Power system transient analysis : theory and practice using simulation programs (ATP-EMTP) / Eiichi Haginomori, Tadashi Koshiduka, Junichi Arai, Hisatochi Ikeda.
      pages   cm
   Includes bibliographical references and index.
   ISBN 978-1-118-73753-8 (cloth)
1. Transients (Electricity)   I. Koshiduka, Tadashi, author.   II. Arai, Junichi, 1932- author.   III. Ikeda, Hisatochi, author.
IV. Title.
   TK3226.P684 2016
   621.319′21–dc23
                                                    2015035756

A catalogue record for this book is available from the British Library.

Cover Image: Casanowe/iStockphoto

Set in 10/12pt Times by SPi Global, Pondicherry, India

1   2016

Contents

Preface

The development of the EMTP (Electro-Magnetic Transient Program) has contributed to a revolution in analysis of switching phenomena and insulation coordination, which are critical issues in modern electric power systems. The authors of this book have been engaged in the development of Japan's electric power system, which is one of the most reliable in the world, as engineers of research and development and in universities for 30–50 years. In their careers, they have used EMTP for solving problems. The contents of this book come from their experiences. Although fundamental examples are displayed, they will definitely be practical for existing power systems.

Some of the contents of the book have been used to teach students in universities and engineers in industry. Those students and engineers all gained a splendid skill that proved useful in their jobs. Electric supply companies and manufacturers need skilled engineers; without them, the modern electric power system cannot operate reliably and safely.

The electric power system is changing rapidly and will change in the future both to cope with the growth of electricity demand and to keep the sustainability of modern society. Designers of today's complicated system configurations and operations need the knowledge in this book more than ever before.

The authors strongly hope that young engineers in the field study this book and use it to contribute to society's future.

Part I

Standard Course-Fundamentals and Typical Phenomena

1

Fundamentals of EMTP

The Electromagnetic Transients Program (EMTP) is a powerful analysis tool for circuit phenomena in power systems. Both steady state voltage and current distribution in the fundamental frequency and surge phenomena in a high-frequency region can be solved using EMTP. Selection of suitable models and appropriate parameters is required for getting correct results. Many comparisons of calculation results and actual recorded data are carried out, and accuracy of EMTP is discussed. Through such applications, EMTP is used widely in the world. EMTP can treat not only main equipment but also control functions. ATP-EMTP is a program that came from EMTP. After ATPDraw (which provides an easy, simple, and powerful graphical user interface) was developed, ATP-EMTP was able to expand its user ability.

1.1 Function and Composition of EMTP

Built-in models in EMTP are listed in Tables 1.1 and 1.2. Table 1.1 shows a main circuit model and Table 1.2 shows a control model. There are two ways to simulate control; one is TACS (Transient Analysis of Control Systems) and the other is MODELS. MODELS is a flexible modeling language and permits more complex calculations than TACS. All statements in MODELS must be written by the user. MODELS is not covered in this book, but TACS is explained for representing control.

1.1.1 Lumped Parameter RLC

The Series RLC Branch model is prepared for representing power system circuits. Load, shunt reactor, shunt capacitor, filter, and other lumped parameter components are represented using this model.

Power System Transient Analysis: Theory and Practice using Simulation Programs (ATP-EMTP), First Edition.
Eiichi Haginomori, Tadashi Koshiduka, Junichi Arai, and Hisatochi Ikeda.
© 2016 John Wiley & Sons, Ltd. Published 2016 by John Wiley & Sons, Ltd.
Companion website: www.wiley.com/go/haginomori_Ikeda/power

Table 1.1 Main circuit model.

Main Circuit Equipment	Built-in Model
Lumped parameter RLC	Series RLC branch
Transmission line, cable	Mutually coupled RLC element, Multiphase PI equivalent (Type 1, 2, 3)
	Distributed parameter line with lumped R (Type-1, -2, -3)
	Frequency dependent distributed parameter line, JMARTI (Type-1, -2, -3)
	Frequency dependent distributed parameter line, SEMLYEN (Type-1)
Transformer	Single-phase saturable transformer
	Three-phase saturable transformer
	Three-phase three-leg core-type transformer
	Mutually coupled RL element (Type 51, 52)
Nonlinear element	Multiphase time varying resistance (Type 91)
	True nonlinear inductance (Type 93)
	Pseudo nonlinear hysteretic inductor (Type 96)
	Staircase time varying resistance (Type 97)
	Pseudo nonlinear inductor (Type 98)
	Pseudo nonlinear resistance (Type 99)
	TACS controlled resistance for arc model (Type 91)
Arrester	Multiphase time-varying resistance (Type 91)
	Exponential ZnO (Type 92)
	Multiphase piecewise linear resistance with flashover (Type 92)
Switch	Time-controlled switch
	Voltage-controlled switch
	Statistical switch
	Measuring switch
TACS controlled switch	Diode, thyristor (Type 11)
	Purely TACS-controlled switch (Type 13)
Voltage source, current source	Empirical data source (Type 1–9)
	Step function (Type 11)
	Ramp function (Type 12)
	Two slopes ramp function (Type 13)
	Sinusoidal function (Type 14)
	CIGRE surge model (Type 15)
	Simplified HVDC converter (Type 16)
	Ungrounded voltage source (Type 18)
	TACS controlled source (Type 60)
Generator	Three-phase synchronous machine (Type 58, 59)
	Universal machine module (Type 19)
Rotating machine	Universal machine module (Type 19)
Control	TACS
	MODELS

1.1.2 Transmission Line

The multiphase PI-equivalent circuit model, Type 1, 2, and 3, is used as a simple line model. It has mutual coupling inductors and is applicable to a transposed or nontransposed three-phase transmission line.

Table 1.2 Control model.

Control Element	Built-in Function in TACS
Transfer function	$\dfrac{K}{s}, Ks, \dfrac{K}{1+Ts}, \dfrac{Ks}{1+Ts},$ $G\dfrac{1+N_1 s+N_2 s^2+\cdots+N_7 s^7}{1+D_1 s+D_2 s^2+\cdots+D_7 s^7}$
Devices	Frequency sensor (50) Relay operated switch (51) Level triggered switch (52) Transport delay (53) Pulse transport delay (54) Digitizer (55) Point-by-point nonlinear (56) Time sequence switch (57) Controlled integrator (58) Simple derivative (59) Input-If selector (60) Signal selector (61) Sample and track (62) Instantaneous min/max (63) Min/max tracking (64) Accumulator and counter (65) RMS meter (66)
Algebraic and logical expression	$+, -, *, /,$ AND, OR, NOT, EQ, GE, SIN, COS, TAN, ASIN, ACOS, ATAN, LOG, LOG10, EXP, SQRT, ABS Free format FORTRAN
Signal source	DC level (Type 11) Sinusoidal signal (Type 14) Pulse (Type 23) Ramp (Type 24)
Input signal from main circuit	Node voltage (Type 90) Switch current (Type 91) Synchronous machine internal signal (Type 92) Switch state (Type 93)
Output signal to main circuit	On/off signal for TACS-controlled switch Signal for TACS-controlled source Torque and field voltage signals for synchronous machine

The distributed parameter line model with lumped resistance, Type-1, -2, and -3, consists of a lossless distributed parameter line model and constant resistances. The resistance is inserted into the lossless line in the mode. Normally the resistance corresponding to the fundamental frequency is used, then this model is applicable to phenomena from the fundamental frequency to the harmonic frequency, in the 1–2 kHz region.

The frequency-dependent distributed parameter line model developed by J. Marti, Semlyen, takes into account line losses at high frequency, even in an untransposed line. It enables the production of detailed and precise simulation for surge analysis. The required data for use of the model can be obtained using support routine Line Constants or Cable Constants, explained later. Height of transmission line tower, conductor configuration, and necessary data are inputted to the support routine, and the input data for EMTP are calculated by the support routine. Both cables and overhead lines are treated by these support routines.

1.1.3   Transformer

A single-phase saturable transformer model is a basic component that permits a multiwinding configuration. The two- or three-winding model is used in many study cases. A pseudo nonlinear inductor is included in this model for saturation characteristics. Input data are resistance and inductance of each winding. A three-phase saturable transformer model also is prepared. The three-phase three-leg transformer is applied for a core type transformer that has a path for air gap flux generated by a zero sequence component. When a hysteresis characteristic is desired, the pseudo nonlinear hysteretic inductor, Type 96, should be used instead of the incorporated pseudo nonlinear inductor. In such a case, the Type 96 branch will be connected outside of the transformer model. The mutually coupled RL element is used for representing a multiwinding transformer; however, self and mutual inductances of all windings are required for input data. This is used for transition voltage analysis in the transformer, which requires a multiwinding model.

1.1.4   Nonlinear Element

True nonlinear inductance, Type 93, has a limit on the number of elements one circuit can hold. When the true nonlinear is included, an iterative convergence calculation is carried out at each time step. Therefore, one element is permitted in one circuit. If more than two elements are needed, these elements must be in separate circuits or be separated by a distributed parameter line. The distributed parameter line separates the network internally as explained in the next section; it is a marked advantage of the EMTP calculation algorithm.

Pseudo-nonlinear elements are prepared that can be used without such constraints. An iterative convergence calculation is not applied for the pseudo nonlinear element, but a simple method is applied. That is, after one time step is calculated, a new value on the nonlinear characteristic curve is adopted for the next time step. Then if the pseudo-nonlinear element is used, a small time step must be selected, suppressing a larger change of voltage or current in the circuit during one time step. The pseudo-nonlinear reactor, Type 98, is the same as the element included in the saturable transformer model. A residual flux in an iron core is simulated by use of the pseudo-nonlinear hysteretic inductor, Type 96.

For use of TACS controlled resistance for the arc model, Type 91, the arc equation must be composed by TACS functions.

1.1.5   Arrester

In the model Type 92, two models are available: one is the exponential ZnO and the other is the multiphase piecewise linear resistance with flashover. The pseudo-nonlinear resistance is also used as an arrester.

1.1.6 Switch

A time-controlled switch is used for normal open/close operation or fault application. The open action is completed after the current crosses the zero point. A voltage-controlled switch is used as a flashover switch or gap. A statistical switch is used for statistical overvoltage studies.

A measuring switch is always closed, along with current value, though the switch is transferred to TACS for control. The TACS-controlled switch, Type 11, simulates a diode without a firing signal or thyristor with a firing signal, as defined in the TACS controller. A purely TACS-controlled switch, Type 13, closes when the open/close signal becomes 1 and opens when the signal becomes 0, even if the current is flowing. The IGBT (insulated gate bipolar transistor) or self-extinguishing power electronics element is simulated by this switch.

1.1.7 Voltage and Current Sources

Many pattern sources are available and a combination of these sources is applicable.

Sinusoidal function, Type 14, is used for a 50 or 60 Hz power source. If the start time of the source, T-start, is specified in negative, EMTP calculates steady state condition and sets initial values of voltage and current to all branches. The ungrounded voltage source consists of voltage source and ideal transformer without grounding on the circuit side. The TACS-controlled source, Type 60, transfers the calculated signal in TACS to the main circuit as a source.

1.1.8 Generator and Rotating Machine

The three-phase synchronous machines, Type 58 and 59, are modeled by Park equations and permit transient calculations. Three-phase circuits in the machine are assumed to be balanced circuits. Values of internal variables of the machine can be transferred to TACS, and torque and filed voltage can be connected from TACS as input signals for the machine. In this model, a mechanical system of shaft with turbines and generators represented by a mass-spring equivalent equation is included and it permits analysis of sub-synchronous resonance phenomena.

The universal machine module, Type 19, is used for modeling of an induction machine or DC machine.

1.1.9 Control

TACS simulates a control part. Input signals for TACS are node voltages, switch currents, internal variables of the rotating machine, and switch status. Output signals from TACS are the on/off signal for the TACS-controlled switch and torque and field voltage for the synchronous machine. Sufficient signal sources, transfer functions, many devices, and algebraic expressions have been prepared, and free-format FORTRAN expression is permitted in addition. Only TACS calculation without the main circuit is accepted.

1.1.10 Support Routines

Support routines are listed in Table 1.3. These support routines are included in EMTP. In the first step the support routine is used and calculated output is obtained. Second, the obtained data are used as input data for EMTP calculation.

Table 1.3 Support routine.

Support Routine	Function
Cable Constants, Line Constants, Cable Parameters	Calculation of data for frequency-dependent distributed parameter line for overhead line and cable from geometric data and resistivity of the earth
Xformer, Bctran	Calculation of self and mutual inductance of transformer windings from capacity and percentage impedance
Saturation	Calculation of peak value saturation curve from RMS saturation data
Hysteresis	Calculation of hysteresis curve from RMS saturation data

1.2   Features of the Calculation Method

The trapezoidal rule is applied in EMTP for numerical integration [1–3]. A simultaneous differential equation is converted to a simultaneous equation with real number coefficient by the trapezoidal rule. The circuit is represented by a nodal admittance equation. The time step for simulation is fixed and ranges from $t=0$ [s] to T-max [s].

1.2.1   Formulation of the Main Circuit

1.2.1.1   Inductance

Figure 1.1 shows an inductance L between node k and m. The basic equation for this circuit is Equation (1.1).

$$e_k - e_m = L\frac{di_{km}}{dt} \tag{1.1}$$

i_{km} at t is obtained by integration from $t - \Delta t$,

$$i_{km}(t) = i_{km}(t - \Delta t) + \frac{1}{L}\int_{t-\Delta t}^{t}(e_k - e_m)dt \tag{1.2}$$

The trapezoidal rule is applied to a time function of f to get area ΔS as shown in Figure 1.2. We get Equation (1.3).

$$\Delta S = \int_{t-\Delta t}^{t} f(t)dt$$
$$\Delta S = \frac{\Delta t}{2}\big(f(t) + f(t - \Delta t)\big) \tag{1.3}$$

Here $f(t) = e_k(t) - e_m(t)$ is substituted and Equation (1.2) is replaced by Equations (1.4) and (1.5).

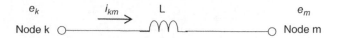

Figure 1.1 Inductance.

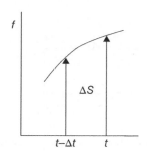

Figure 1.2 Function f and area ΔS.

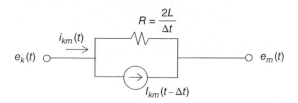

Figure 1.3 Equivalent circuit of inductance.

$$i_{km}(t) = \frac{\Delta t}{2L}\left(e_k(t) - e_m(t)\right) + I_{km}(t - \Delta t) \tag{1.4}$$

$$I_{km}(t - \Delta t) = i_{km}(t - \Delta t) + \frac{\Delta t}{2L}\left(e_k(t - \Delta t) - e_m(t - \Delta t)\right) \tag{1.5}$$

Equation (1.4) is represented by Figure 1.3.

Equation (1.5) is the value of the previous step and is a known value at calculation of time t. Figure 1.3 shows that the inductance is represented by parallel connection of an equivalent resistance R and the known current source. Resistance R is calculated once before time step calculation.

1.2.1.2 Capacitance

Capacitance C between nodes k and m is shown in Figure 1.4. The basic equation for this circuit is Equation (1.6).

$$i_{km} = C\frac{d\left(e_k - e_m\right)}{dt} \tag{1.6}$$

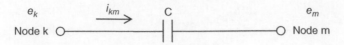

Figure 1.4   Capacitance.

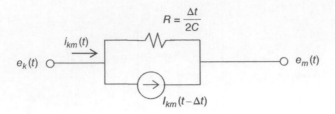

Figure 1.5   Equivalent circuit of capacitance.

By applying the trapezoidal rule to Equation (1.6), Equations (1.7) and (1.8) are obtained and equivalent circuit is shown as in Figure 1.5; that means the capacitance is represented by an equivalent resistance R and a known current source.

$$i_{km}(t) = \frac{2C}{\Delta t}\left(e_k(t) - e_m(t)\right) + I_{km}(t - \Delta t) \tag{1.7}$$

$$I_{km}(t - \Delta t) = -i_{km}(t - \Delta t) - \frac{2C}{\Delta t}\left(e_k(t - \Delta t) - e_m(t - \Delta t)\right) \tag{1.8}$$

1.2.1.3   Resistance

The resistance shown in Figure 1.6 is represented as it appears.

1.2.1.4   Distributed Parameter Line

The distributed parameter line connecting node k and node m is shown in Figure 1.7.

If resistance is ignored, relationships between voltage and current as functions of distance and time are described in differential equations, Equation (1.9).

$$-\frac{\partial e}{\partial x} = L'\left(\frac{\partial i}{\partial t}\right)$$
$$-\frac{\partial i}{\partial x} = C'\left(\frac{\partial e}{\partial t}\right) \tag{1.9}$$

where L' and C' are inductance and capacitance per unit length, respectively. The solution is shown in Equation (1.10),

$$e_k(t - \tau) + Z * i_{km}(t - \tau) = e_m(t) - Z * i_{mk}(t)$$
$$e_k(t) - Z * i_{km}(t) = e_m(t - \tau) + Z * i_{mk}(t - \tau) \tag{1.10}$$
$$Z = \sqrt{\frac{L'}{C'}}, \quad v = \frac{1}{\sqrt{L' * C'}}, \tau = \frac{1}{v}$$

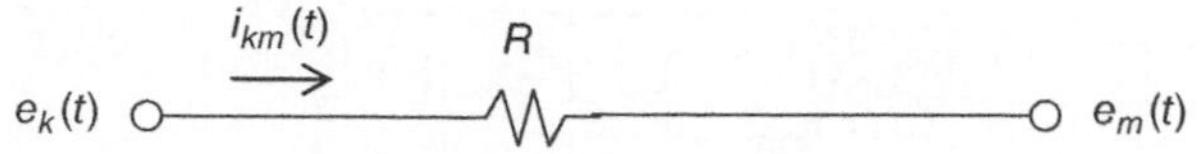

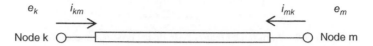

Figure 1.6   Resistance circuit.

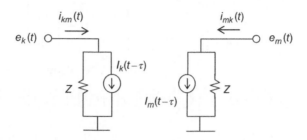

Figure 1.7   Distributed parameter line.

Figure 1.8   Equivalent circuit for distributed parameter line.

where Z is surge impedance, v is propagation velocity, and τ is travel time. Equation (1.10) can be represented in Figure 1.8.

At node k, voltage and current are expressed by Equation (1.11). This means the current $i_{km}(t)$ is represented by voltage at self node k and a known current before travel time τ. As a result, the two nodes can be treated as separated circuits.

$$i_{km}(t)=\frac{1}{Z}e_{k}(t)+I_{k}(t-\tau)$$

$$I_{k}(t-\tau)=-\frac{1}{Z}e_{m}(t-\tau)-i_{mk}(t-\tau)$$

(1.11)

1.2.1.5   Nodal Equation

The nodal equation of the circuit is formulated in Equation (1.12) by applying the trapezoidal rule.

$$[Y]*[e(t)]=[i(t)]-[I]$$

(1.12)

where Y = node conductance matrix (real value), $i(t)$ = injection current vector, and I = known current vector.

Equation (1.12) is represented as Equation (1.13) by dividing it into unknown and known values. Finally, the unknown value is solved as in Equation (1.14).

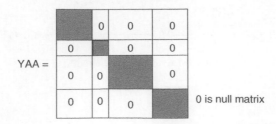

Figure 1.9 Admittance matrix with distributed parameter lines.

Unknown   Known

$$\begin{bmatrix} Y_{AA} & Y_{AB} \\ Y_{BA} & Y_{BB} \end{bmatrix}\begin{bmatrix} e_A(t) \\ e_B(t) \end{bmatrix} = \begin{bmatrix} i_A(t) \\ i_B(t) \end{bmatrix} - \begin{bmatrix} I_A \\ I_B \end{bmatrix} \qquad (1.13)$$

Known   Unknown

$$\left[e_A(t) \right] = \left[Y_{AA} \right]^{-1} * \left\{ \left[i_A(t) \right] - \left[I_A \right] - \left[Y_{AB} \right] * \left[e_B(t) \right] \right\} \qquad (1.14)$$

In EMTP voltage, $e_A(t)$ is calculated at each time step until T-max is reached.

If a distributed parameter line is used in the circuit, the admittance matrix Y_{AA} is divided into a small size matrix as shown in Figure 1.9, due to Figure 1.8. It contributes a short computation time and error reduction.

1.2.2   Calculation in TACS

The trapezoidal rule is also applied in TACS. A general transfer function, $G(s)$ of Equation (1.15), is taken for explanation.

$$X(s) = G(s) * U(s)$$

$$G(s) = \frac{N_0 + N_1 s + N_2 s^2 + \cdots + N_m s^m}{D_0 + D_1 s + D_2 s^2 + \cdots + D_n s^n} \qquad (1.15)$$

U is input, X is output, and s is a Laplace operator.

Laplace operator s is replaced by d/dt for transient analysis, then Equation (1.15) is represented by differential Equation (1.16).

$$D_0 x + D_1 \frac{dx}{dt} + D_2 \frac{d^2 x}{dt^2} + \cdots + D_n \frac{d^n x}{dt^n} = N_0 u + N_1 \frac{du}{dt} + N_2 \frac{d^2 u}{dt^2} + \cdots + N_m \frac{d^m u}{dt^m} \qquad (1.16)$$

New variables are introduced as follows:

$$x_1 = \frac{dx}{dt}, x_2 = \frac{dx_1}{dt}, \cdots, x_n = \frac{dx_{(n-1)}}{dt}$$

$$u_1 = \frac{du}{dt}, u_2 = \frac{du_1}{dt}, \cdots, u_n = \frac{du_{(m-1)}}{dt}$$

The trapezoidal rule is applied for $x_1 = \frac{dx}{dt}$,

$$x_1 = \frac{2}{\Delta t} x(t) - \left\{ x_1(t - \Delta t) + \frac{2}{\Delta t} x(t - \Delta t) \right\}$$

The second term on the right hand side is the known value. Finally, Equation (1.15) is represented by simultaneous linear equations.

$$c * x(t) = d * u(t) + Hist(t - \Delta t) \tag{1.17}$$

where c and d are coefficients. They are calculated uniquely by time step Δt and parameters of transfer function. The calculation of these coefficients is required once before transient calculation.

1.2.3   Features of EMTP

1.2.3.1   Relationship between the Main Circuit and TACS

Although the main circuit and TACS part must be solved essentially simultaneously, EMTP calculates them independently [4]. The main circuit at time t is calculated initially. Voltage and current signals are transferred to TACS and calculation of TACS is carried out. The output of TACS is used in the main circuit calculation of the next time, $t + \Delta t$. The output of TACS is on/off pulse signal for TACS-controlled switch or exciter voltage for synchronous generator. In most cases, the controller has a delay at the input and output stages, so selection of a reasonably small time step will make the error negligible. If a large time step is selected, attention should be given to the calculation error.

1.2.3.2   Initial Setting

Initial values on all branches are set automatically if a negative T-start of the sinusoidal source is specified. EMTP calculates the steady state condition by complex plane and the value of the real part is set to each branch. The steady state calculation permits only one frequency. In TACS, the initial DC value can be inputted by the user.

1.2.3.3   Nonlinear Branch

Only one true nonlinear branch is accepted due to performing the iterative convergence calculation. There is no such restriction for a pseudo-nonlinear branch.

1.2.3.4   Floating Circuit

A floating circuit must be avoided due to calculation error at the inverse calculation of the admittance matrix. This is measured by connecting the stray capacitance to the ground.

1.2.3.5 Calculation Order in TACS

Calculation order of control elements is determined automatically. When the device in Table 1.2 is used, EMTP cannot determine its order. The user must specify its order by indicating the place, in input side, in output side, or internal position between transfer functions.

1.2.3.6 Switch and Apparent Oscillation

In normal use of the time switch when the open order is given to the switch, the switch memorizes the current direction and opens after the current changes the sign, that is, from plus to minus, or from minus to plus. EMTP adopts the fixed time step calculation, so then the switch current is not zero at the opened time. Due to this algorithm, apparent oscillation appears on the voltage at the terminal of inductance, as shown in Figure 1.10a, b. Figure 1.10a shows the interruption of pure inductance current and voltage at node V1. At node V1 there is no branch to the ground and apparent voltage oscillation is obtained. This oscillation appears at each time step. In an actual system there is no such condition; part of the branch exists as a stray capacitor. In Figure 1.11a, b a small capacitance, 10 μF, is connected at node V2, and the apparent oscillation disappears. This problem that causes current oscillation when pure capacitance is closed by the switch can be solved by adding reactance in series.

1.2.3.7 ATPDraw

ATPDraw is a graphical preprocessor for ATP-EMTP [5], and it allows execution of ATP-EMTP and PLOTXY. Figure 1.12 shows a simple outline and relating files for normal use.

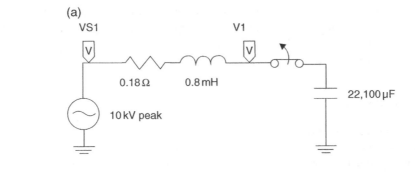

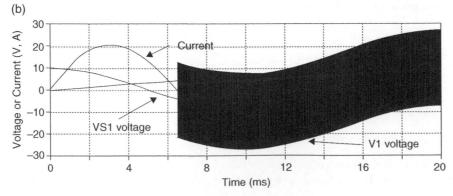

Figure 1.10 Reactor current interruption. (a) Circuit. (b) Current and voltages.

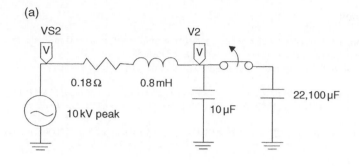

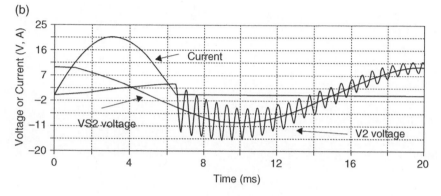

Figure 1.11 Reactor current interruption with capacitor modification. (a) Circuit. (b) Current and voltages.

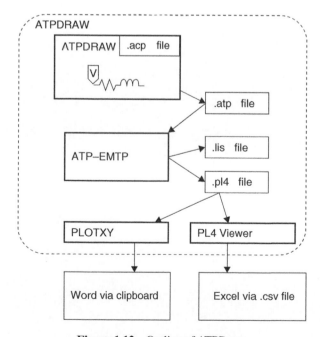

Figure 1.12 Outline of ATPDraw.

References

[1] H. W. Dommel (1969) Digital computer solution of electromagnetic transients in single- and multiphase network, *IEEE Transactions on Power Apparatus and Systems*, **PAS-88**, 4, 388–399.

[2] H. W. Dommel, W. S. Meyer (1974) Computation of electromagnetic transients, *Proceeding of the IEEE*, **62** (7), 983–993.

[3] H.W. Dommel (1986) *Electromagnetic Transients Program Reference Manual (EMTP Theory Book)*, BPA.

[4] W. Scott Meyer, T.-H. Liu (1992) *Alternative Transients Program (ATP) Rule Book*, Canadian/American EMTP User Group.

[5] L. Prikler, H. K. Hoidalen (2002) *ATPDRAW Version 3.5 for Windows 9x/NT/2000/XP Users' Manual*, SINTEF.

2

Modeling of System Components

2.1 Overhead Transmission Lines and Underground Cables

2.1.1 Overhead Transmission Line—Line Constants

2.1.1.1 General

2.1.1.1.1 Inductance of Single Conductor over the Earth

When a single conductor is located over the Earth's surface, the magnetic images are as shown in Figure 2.1 and self-inductance is written as in Equation (2.1). Table 2.1 shows the self-inductances in the case of changing H_e. The self-inductance increases with H_e.

$$L = \frac{\mu_0}{2\pi}\left(1 + 2\ln\frac{2H_e}{r}\right) \tag{2.1}$$

2.1.1.1.2 Capacitance of Single Conductor over the Earth

A conductor with a radius r (m) is located at h (m) high over the Earth and has an electric charge $+q$ (c) per unit length, as shown in Figure 2.2. Generally, the permittivity of Earth, εe, is quite large compared with the permittivity of air, εa. The electric field lines from the conductor will flow into the Earth vertically. The voltage distribution of Earth's surface will be flat. We can treat the Earth as a conductor. The electric field in the air will be treated as shown in Figure 2.3.

The voltage between two conductors is expressed by Equation (2.2).

$$v_{11} = \frac{q}{\pi\varepsilon_0}\ln\frac{2h}{r} \tag{2.2}$$

Power System Transient Analysis: Theory and Practice using Simulation Programs (ATP-EMTP), First Edition.
Eiichi Haginomori, Tadashi Koshiduka, Junichi Arai, and Hisatochi Ikeda.
© 2016 John Wiley & Sons, Ltd. Published 2016 by John Wiley & Sons, Ltd.
Companion website: www.wiley.com/go/haginomori_Ikeda/power

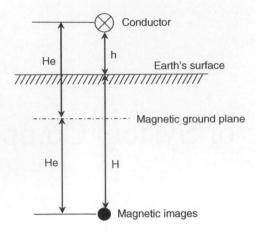

Figure 2.1   Depth of Earth return.

Table 2.1   Inductance of single conductor over the Earth.

H_e (m)	L (mH/km)
10	1.55
50	1.87
100	2.01
300	2.23
500	2.33
1000	2.47

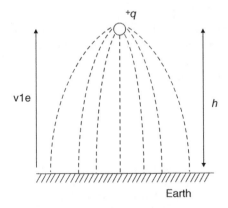

Figure 2.2   Electric field lines from the conductor.

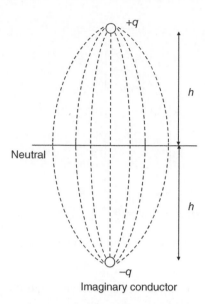

Figure 2.3 Electric field in the air.

So, the voltage between the conductor and the earth is expressed by Equation (2.3).

$$v_{1e} = \frac{v_{11}}{2} = \frac{q}{2\pi\varepsilon_0} \ln \frac{2h}{r} \tag{2.3}$$

The capacitance between the conductor and the earth C_{1e} is expressed by Equation (2.4).

$$C_{1e} = \frac{q}{v_{1e}} = \frac{2\pi\varepsilon_0}{\ln \dfrac{2h}{r}} \tag{2.4}$$

In case of the transposed line, the positive sequence capacitance is expressed by Equation (2.5).

$$C_1 = \frac{2\pi\varepsilon_0}{\ln \dfrac{D}{R'}} \tag{2.5}$$

That capacitance is decided by the equivalent distance D between phases and the equivalent radius of conductor R'.

The height of the line does not affect the capacitance.

The zero sequence capacitance is expressed by Equation (2.6).

$$C_0 = \frac{2\pi\varepsilon_0}{\ln \dfrac{8h^3}{R'D^2}} \tag{2.6}$$

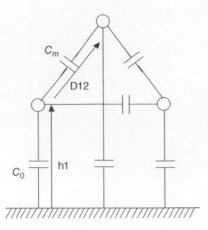

Figure 2.4 Capacitances between three-phase conductors.

The relation between positive and zero sequence capacitance is written in Equation (2.7).

$$C_1 = C_0 + 3C_m \tag{2.7}$$

Capacitance between phases is expressed by Equation (2.8) (Figure 2.4).

$$C_m = \frac{1}{3}(C_1 - C_0) \tag{2.8}$$

2.1.1.1.2.1 Capacitance Matrix

Here, we imagine three phases conductor over the Earth. The voltages V_1, V_2, and V_3 are a function of the line charges q_1, q_2, and q_3, as shown in Equation (2.9). The p_{ls} is Maxwell's potential coefficient.

$$\begin{bmatrix} V_1 \\ V_2 \\ V_3 \end{bmatrix} = \begin{bmatrix} p_{11} & p_{12} & p_{13} \\ p_{21} & p_{22} & p_{23} \\ p_{31} & p_{32} & p_{33} \end{bmatrix} \begin{bmatrix} q_1 \\ q_2 \\ q_3 \end{bmatrix} \tag{2.9}$$

In general cases,

$$[v] = [p][q], \tag{2.10}$$

where

$$p_{ii} = \frac{1}{2\pi\varepsilon_0} \ln \frac{2h_i}{r_i} (\mathrm{m/F}) \quad p_{ik} = p_{ki} = \frac{1}{2\pi\varepsilon_0} \ln \frac{D_{ik}}{d_{ik}} (\mathrm{m/F}).$$

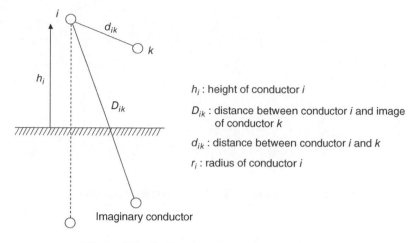

Figure 2.5   Explanation of mutual capacitance C_{ik}.

The capacitance matrix is derived as an inverse matrix of Equation (2.10):

$$[q]=[C][v] \quad [C]=[p]^{-1} \quad \begin{bmatrix} q_1 \\ q_2 \\ q_3 \end{bmatrix} = \begin{bmatrix} k_{11} & k_{12} & k_{13} \\ k_{21} & k_{22} & k_{23} \\ k_{31} & k_{32} & k_{33} \end{bmatrix} \begin{bmatrix} V_1 \\ V_2 \\ V_3 \end{bmatrix} \tag{2.11}$$

$$\begin{bmatrix} p_{11} & p_{12} & p_{13} \\ p_{21} & p_{22} & p_{23} \\ p_{31} & p_{32} & p_{33} \end{bmatrix}^{-1} = \begin{bmatrix} k_{11} & k_{12} & k_{13} \\ k_{21} & k_{22} & k_{23} \\ k_{31} & k_{32} & k_{33} \end{bmatrix}. \tag{2.12}$$

The q_1, q_2, and q_3 are transformed as follows:

$$
\begin{aligned}
q_1 &= k_{11}\cdot V1 + k_{12}\cdot V2 + k_{13}\cdot V3 = \left(k_{11}+k_{12}+k_{13}\right)\cdot V_1 + \left(-k_{12}\right)\cdot\left(V_1-V_2\right)+\left(-k_{13}\right)\cdot\left(V_1-V_3\right) \\
q_2 &= k_{21}\cdot V1 + k_{22}\cdot V2 + k_{23}\cdot V3 = \left(-k_{21}\right)\cdot\left(V_2-V_1\right)+\left(k_{21}+k_{22}+k_{23}\right)\cdot V_2 + \left(-k_{23}\right)\cdot\left(V_2-V_3\right) \\
q_3 &= k_{31}\cdot V1 + k_{32}\cdot V2 + k_{33}\cdot V3 = \left(-k_{31}\right)\cdot\left(V_3-V_1\right)+\left(-k_{32}\right)\cdot\left(V_3-V_2\right)+\left(k_{31}+k_{32}+k_{33}\right)\cdot V_3.
\end{aligned}
\tag{2.13}
$$

The coefficient of V_1, V_2, and V_3 is the capacitance from the conductor to the ground, and they are written as C_{11}, C_{12}, and C_{13}. Another coefficient is mutual capacitance C_{ik} (Figure 2.5):

$$
\begin{array}{ll}
C_{11}=k_{11}+k_{12}+k_{13} & C_{12}=C_{21}=-k_{12} \\
C_{22}=k_{21}+k_{22}+k_{23} & C_{23}=C_{32}=-k_{23} \\
C_{33}=k_{31}+k_{32}+k_{33} & C_{31}=C_{13}=-k_{31}
\end{array}
\tag{2.14}
$$

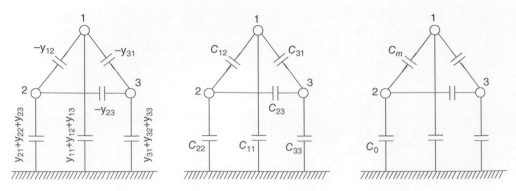

Figure 2.6 Explanation of C_0, C_m at the three-phase conductor.

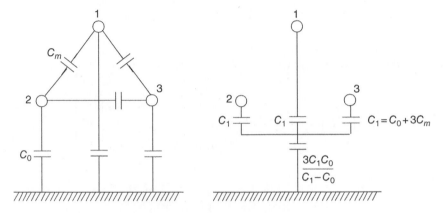

Figure 2.7 Explanation of C_1 (positive sequence capacitance).

In case of the transposed line (Figures 2.6 and 2.7),

$$C_{11} = C_{22} = C_{33} = C_0, \ \ C_{12} = C_{23} = C_{31} = C_\mathrm{m}. \tag{2.15}$$

2.1.1.1.3 Elimination of Ground Wire (Figure 2.8)

In Figure 2.4, the relation between voltages, impedances, and currents is shown as Equation (2.16).

$$\begin{bmatrix} V_a \\ V_b \\ V_c \\ V_v \\ V_w \end{bmatrix} = \begin{bmatrix} Z_{aa} & Z_{ab} & Z_{ac} & Z_{av} & Z_{aw} \\ Z_{ba} & Z_{bb} & Z_{bc} & Z_{bv} & Z_{bw} \\ Z_{ca} & Z_{cb} & Z_{cc} & Z_{cv} & Z_{cw} \\ Z_{va} & Z_{vb} & Z_{vc} & Z_{vv} & Z_{vw} \\ Z_{wa} & Z_{wb} & Z_{wc} & Z_{wv} & Z_{ww} \end{bmatrix} \begin{bmatrix} I_a \\ I_b \\ I_c \\ I_v \\ I_w \end{bmatrix} + \begin{bmatrix} V_a' \\ V_b' \\ V_c' \\ V_v' \\ V_w' \end{bmatrix} \tag{2.16}$$

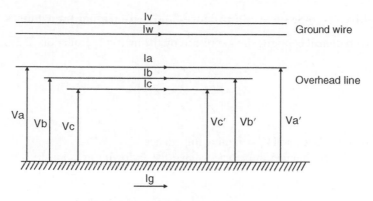

Figure 2.8 Elimination for ground wire.

$$
\begin{bmatrix} V_a \\ V_b \\ V_c \\ V_v \\ V_w \end{bmatrix} - \begin{bmatrix} V_a' \\ V_b' \\ V_c' \\ V_v' \\ V_w' \end{bmatrix} = \begin{bmatrix} \Delta V_a \\ \Delta V_b \\ \Delta V_c \\ \Delta V_v \\ \Delta V_w \end{bmatrix} = \begin{bmatrix} Z_{aa} & Z_{ab} & Z_{ac} & Z_{av} & Z_{aw} \\ Z_{ba} & Z_{bb} & Z_{bc} & Z_{bv} & Z_{bw} \\ Z_{ca} & Z_{cb} & Z_{cc} & Z_{cv} & Z_{cw} \\ Z_{va} & Z_{vb} & Z_{vc} & Z_{vv} & Z_{vw} \\ Z_{wa} & Z_{wb} & Z_{wc} & Z_{wv} & Z_{ww} \end{bmatrix} \begin{bmatrix} I_a \\ I_b \\ I_c \\ I_v \\ I_w \end{bmatrix}
\tag{2.17}
$$

When we assume that the ground wire voltage is zero in the alternative frequency, we derive $\Delta V_v = \Delta V_w = 0$.

The lower half of the matrix is expressed by Equation (2.18).

$$
\begin{bmatrix} \Delta V_v \\ \Delta V_w \end{bmatrix} = \begin{bmatrix} 0 \\ 0 \end{bmatrix} = \begin{bmatrix} Z_{va} & Z_{vb} & Z_{vc} \\ Z_{wa} & Z_{wb} & Z_{wc} \end{bmatrix} \begin{bmatrix} I_a \\ I_b \\ I_c \end{bmatrix} + \begin{bmatrix} Z_{vv} & Z_{vw} \\ Z_{wv} & Z_{ww} \end{bmatrix} \begin{bmatrix} I_v \\ I_w \end{bmatrix}
\tag{2.18}
$$

So, we derive this equation:

$$
\begin{bmatrix} I_v \\ I_w \end{bmatrix} = -\begin{bmatrix} Z_{vv} & Z_{vw} \\ Z_{wv} & Z_{ww} \end{bmatrix}^{-1} \begin{bmatrix} Z_{va} & Z_{vb} & Z_{vc} \\ Z_{wa} & Z_{wb} & Z_{wc} \end{bmatrix} \begin{bmatrix} I_a \\ I_b \\ I_c \end{bmatrix}.
\tag{2.19}
$$

The upper half of the matrix will be shown as Equation (2.20).

$$
\begin{bmatrix} \Delta V_a \\ \Delta V_b \\ \Delta V_c \end{bmatrix} = \begin{bmatrix} Z_{aa} & Z_{ab} & Z_{ac} \\ Z_{ba} & Z_{bb} & Z_{bc} \\ Z_{ca} & Z_{cb} & Z_{cc} \end{bmatrix} \begin{bmatrix} I_a \\ I_b \\ I_c \end{bmatrix} + \begin{bmatrix} Z_{av} & Z_{aw} \\ Z_{bv} & Z_{bw} \\ Z_{cv} & Z_{cw} \end{bmatrix} \begin{bmatrix} I_v \\ I_w \end{bmatrix}
\tag{2.20}
$$

To insert Equation (2.19) into Equation (2.20), we eliminate the ground wire.

2.1.1.2 Positive-/Zero-Sequence Impedances and Self-/Mutual Impedances

Impedances with mutual coupling are expressed by the matrix in Equation (2.21),

$$Z = \begin{bmatrix} Z_S & Z_m & Z_m \\ Z_m & Z_S & Z_m \\ Z_m & Z_m & Z_S \end{bmatrix}, \tag{2.21}$$

where Z_s = self-impedance and Z_m = mutual impedance.

Zero- and positive-sequence impedances have the following relation to self- and mutual impedances:

$$\begin{aligned} Z_{POS} &= Z_S - Z_m \\ Z_{ZERO} &= Z_S + 2Z_m, \end{aligned} \tag{2.22}$$

where Z_{POS} = positive-sequence impedance and Z_{ZERO} = zero-sequence impedance.

Inversely, we get the relation expressed by Equation (2.23):

$$\begin{aligned} Z_S &= \frac{1}{3}\left(2Z_{POS} + Z_{ZERO}\right) \\ Z_m &= \frac{1}{3}\left(Z_{ZERO} - Z_{POS}\right). \end{aligned} \tag{2.23}$$

2.1.1.3 Overhead Line Model in EMTP-ATP

A single transmission line is modeled by using lumped elements as shown in Figure 2.9. This is the ideal circuit in case of Δx being infinitesimal.

2.1.1.3.1 Pi Equivalents

A single transmission line is modeled by the four terminal equivalent circuits shown in Figure 2.10.

This model has efficient accuracy when the frequency of calculation phenomena is sufficiently smaller than the inverse of propagation time of the line. Generally, users can use this pi equivalent model under the 100-km line. In case of a greater than 100-km line, users can use this model by connecting in series.

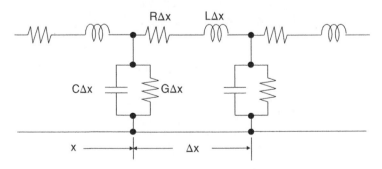

Figure 2.9   Equivalent circuit of a single-phase transmission line.

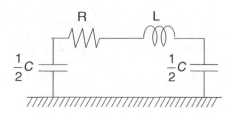

Figure 2.10 Pi equivalent circuit.

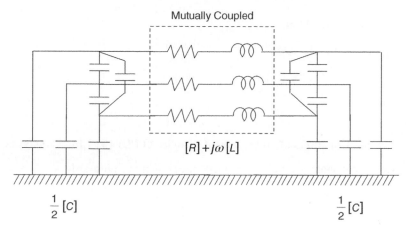

Figure 2.11 Pi equivalent circuit for three-phase lines.

In the three-phase transmission line, an equivalent circuit is modeled by the mutual coupling shown in Figure 2.11.

2.1.1.3.2 *Distortionless Line Modeling*
This is the line model under R=G=0 in Figure 2.9.

The characteristic impedance and propagation velocity are expressed by Equations (2.24) and (2.25):

$$\text{Characteristic impedance} \left(\text{Surge impedance}\right) Z = \sqrt{\frac{L}{C}} \qquad (2.24)$$

$$\text{Propagation velocity } v = \frac{1}{\sqrt{LC}}. \qquad (2.25)$$

2.1.1.3.3 *Lumped-Resistance Line Modeling*
This model shown in Figure 2.12 is a distortionless line model with lumped resistance.

This model is necessary to be used in case of $1/4R \ll Z$.

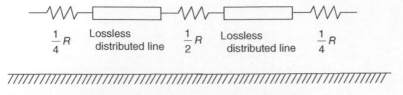

Figure 2.12   Lumped-resistance line model.

2.1.1.3.4   *Frequency-Dependent Line Modeling*

EMTP-ATP has the support programs listed below for calculating line constants.

The line constants and J. Marti models are most popular. These models can treat frequency dependence of the overhead line.

- Line Constants
- J. Marti
- Semlyen
- NODA Setup

2.1.1.4   Example Constants of an Overhead Line in ATPDraw: Line Constants

ATPDraw tries to calculate overhead line constants, shown in Figure 2.13.

Overhead line $410\,mm^2$ Two bundled conductors $S = 12.5\,cm$
            Diameter 2.85 cm, diameter of steel wire 1.05 cm, DC resistance 0.0671 Ω/km
Ground wire $120\,mm^2$ Single conductor
            Diameter 1.75 cm, diameter of steel wire 1.05 cm, DC resistance 0.277 Ω/km

- Selection of LCC (short for Line Constants, Cable Constants, or cable parameters) in the menu, which is shown in Figure 2.14.
- Setting for the Bergeron model, which is shown in Figure 2.15.
- It is necessary to change the frequency for the phenomena.
- With the phenomena at an alternate power frequency, Freq.init is used at 50 or 60 Hz.

In the case of high-frequency phenomena, Freq.init is $c_0/4\ell$, where ℓ = line length and C_0 = light speed. The following are terms used in ATPDraw:

- "Transposed": transposed or not
- "Auto bundling": automatic bundling option that which allows a single conductor
- "Skin effect": frequency-dependent for resistance or not
- "Segmented ground": ground wires are to be treated as being continuous or segmented.

Users should select the proper model from the following list.

- Bergeron: line constants model
- PI: π equivalents
- J. Marti: J. Marti model
- Semlyen: Semlyen model
- Noda: Noda setup model

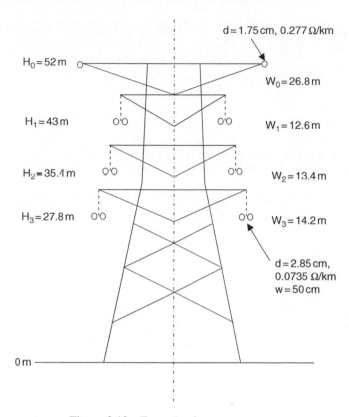

$H_0 = 52\,m$ $d = 1.75\,cm,\ 0.277\,\Omega/km$

$W_0 = 26.8\,m$

$H_1 = 43\,m$ $W_1 = 12.6\,m$

$H_2 = 35.4\,m$ $W_2 = 13.4\,m$

$H_3 = 27.8\,m$ $W_3 = 14.2\,m$

$d = 2.85\,cm,$
$0.0735\ \Omega/km$
$w = 50\,cm$

$0\,m$

Figure 2.13   Example of a tower structure.

Figure 2.14   LCC menu in ATPDraw.

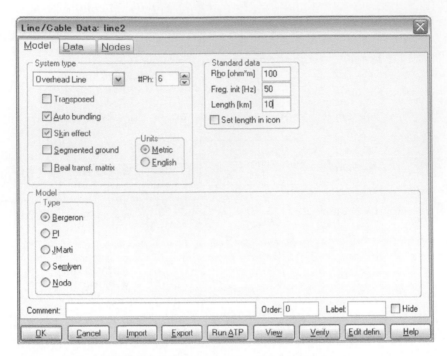

Figure 2.15   Bergeron model in ATPDraw.

Data tag: Setting for geometrical information to calculate the line constants, see Figure 2.16.

Ph.no.: Line number, "0" means a ground wire that is grounded. This means the elimination of ground wire. Otherwise, when the ground wire is treated as a conductor, this should be input by the number except "0."

Horiz: Horizontal distance from center of tower to conductor.

     If the right side is plus, the left side is minus.

Vmid: Vertical height of conductor above ground at mid-span.

     If the distance between each tower is larger than 200 m, users can set this value at Vtower × 2/3.

Separ, Alpha: Distance of each conductor and degree.

In case of the left figure, users should set alpha = 45° (Figure 2.17).

2.1.1.4.1   *Example of Calculated Constants*

In the case of a transposed line, zero- and positive-sequence constants are calculated. The first line shows a zero sequence and the second line shows a positive sequence. The negative sequence is equated to the positive.

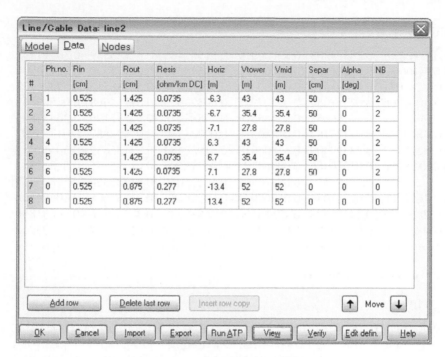

Figure 2.16 Data setting in ATPDraw.

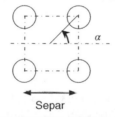

Separ

In the case of the left figure, users should set Alpha = 45°.

Figure 2.17 Explanation of "Separ" and "Alpha."

Nontransposed line, at 50 Hz

$VINTAGE, 1

```
-1IN___AOUT__A      2.82564E-01 9.08011E+02 2.40724E+05-1.00000E+01 1   6
-2IN___BOUT__B      3.82969E-02 3.68089E+02 2.95893E+05-1.00000E+01 1   6
-3IN___COUT__C      4.06809E-02 3.30997E+02 2.95530E+05-1.00000E+01 1   6
-4IN___DOUT__D      3.74537E-02 2.88842E+02 2.96595E+05-1.00000E+01 1   6
-5IN___EOUT__E      3.69859E-02 2.54055E+02 2.96306E+05-1.00000E+01 1   6
-6IN___FOUT__F      3.68752E-02 2.60961E+02 2.96418E+05-1.00000E+01 1   6
```

$VINTAGE, 0

```
 0.33312492  0.28399308 –0.51806919 –0.51982484 –0.37135639 –0.30411451
–0.07046654 –0.01016986  0.00064779  0.00518695 –0.01175546 –0.02818906
 0.32283071  0.47191570 –0.05190306 –0.10261031  0.51788356  0.57826777
 0.01077706  0.02931034 –0.05207941  0.00524663 –0.01754668 –0.00582566
 0.52665125  0.44166653  0.47545596  0.46813048 –0.30532196 –0.26831886
 0.04863027 –0.02473182 –0.01180385  0.00725787 –0.01494887  0.01723907
 0.33312492 –0.28399308 –0.51806919  0.51982484  0.37135639 –0.30411451
–0.07046654  0.01016986  0.00064779 –0.00518695  0.01175546 –0.02818906
 0.32283071 –0.47191570 –0.05190306  0.10261031 –0.51788356  0.57826777
 0.01077706 –0.02931034 –0.05207941 –0.00524663  0.01754668 –0.00582566
 0.52665125 –0.44166653  0.47545596 –0.46813048  0.30532196 –0.26831886
 0.04863027  0.02473182 –0.01180385 –0.00725787  0.01494887  0.01723907
```

Nontransposed line, at 5000 Hz

$VINTAGE, 1

```
–1IN___AOUT__A      7.85679E+00 7.88809E+02 2.71036E+05–1.00000E+01 1   6
–2IN___BOUT__B      2.13788E–01 3.52941E+02 2.98018E+05–1.00000E+01 1   6
–3IN___COUT__C      1.99237E–01 3.36528E+02 2.98771E+05–1.00000E+01 1   6
–4IN___DOUT__D      1.68803E–01 2.91494E+02 2.98994E+05–1.00000E+01 1   6
–5IN___EOUT__E      1.65661E–01 2.53865E+02 2.98905E+05–1.00000E+01 1   6
–6IN___FOUT__F      1.65973E–01 2.59424E+02 2.98924E+05-1.00000E+01 1   6
```

$VINTAGE, 0

```
 0.25078561 –0.23132315 –0.49751501  0.55619157 –0.35553940 –0.35970967
–0.01475895  0.00637573 –0.02152691  0.00248763  0.00039556  0.01923165
 0.31677409 –0.30354557 –0.14947397  0.20745866  0.57540790  0.56479156
–0.00234087 –0.03419151  0.02463887 –0.01613332 –0.00794804  0.00550554
 0.57998573 –0.59395802  0.47823316 –0.38379552 –0.20403499 –0.22548265
 0.01240975  0.01881070 –0.01894511 –0.00680935 –0.02835272 –0.01941282
 0.25078561  0.23132315 –0.49751501 –0.55619157  0.35553940 –0.35970967
–0.01475895 –0.00637573 –0.02152691 –0.00248763 –0.00039556  0.01923165
 0.31677409  0.30354557 –0.14947397 –0.20745866 –0.57540790  0.56479156
–0.00234087  0.03419151  0.02463887  0.01613332  0.00794804  0.00550554
 0.57998573  0.59395802  0.47823316  0.38379552  0.20403499 –0.22548265
 0.01240975 –0.01881070 –0.01894511  0.00680935  0.02835272 –0.01941282
```

For example, Figure 2.18 shows the calculated results for applying these three lines to a 100 V source.

In Figure 2.18, oscillation decays fast at the 5000 Hz line. The difference whether transposed or not is small.

Transposed line, 5000 Hz

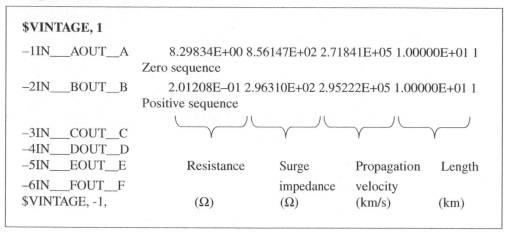

$VINTAGE, 1

–1IN___AOUT__A	8.29834E+00 8.56147E+02 2.71841E+05 1.00000E+01 1			
	Zero sequence			
–2IN___BOUT__B	2.01208E–01 2.96310E+02 2.95222E+05 1.00000E+01 1			
	Positive sequence			

–3IN___COUT__C				
–4IN___DOUT__D				
–5IN___EOUT__E	Resistance	Surge	Propagation	Length
–6IN___FOUT__F		impedance	velocity	
$VINTAGE, -1,	(Ω)	(Ω)	(km/s)	(km)

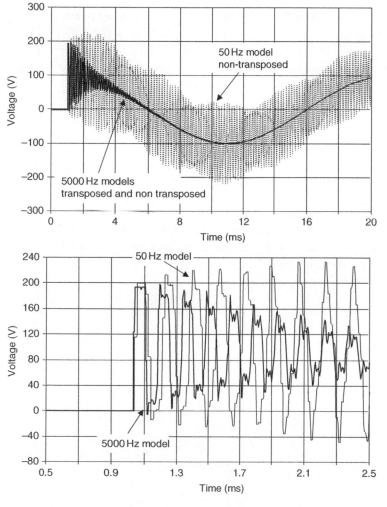

Figure 2.18 Calculated results for making lines.

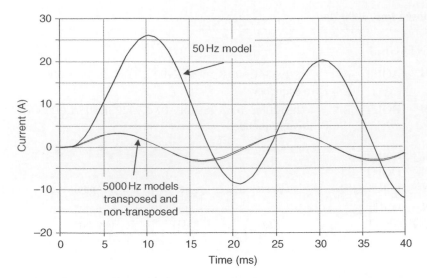

Figure 2.19   Calculated result for short-circuit current.

Figure 2.19 shows the comparison for a short-circuit current. At 5000 Hz, the amplitude of the short-circuit current is different from the 50 Hz line.

Short line fault interruption has alternative frequency phenomena and high-frequency phenomena. At current interruption, the line constant should be treated as an alternative frequency. But transient recovery voltage shows the several 10 kHz. A two-step calculation is necessary for accuracy. In this case, the current injection method is suitable. At first, the user should calculate the short-circuit current for an alternative frequency. Second, the user should inject that current to the high-frequency lines.

The J. Marti model is widely used for frequency-dependent transmission line model. The line constant is calculated as follows.

The J. Marti model setting: the user should select "J. Marti" in the Model window. Data subwindow will be shown. The user should set the frequency at the steady state and calculation frequency for the mode matrix.

For example,

- Freq.matrix: mode matrix frequency, here 5 kHz
- Freq.SS: steady state frequency, here 50 Hz
- Decades: frequency range for constant calculation, here 0.1 10^8 Hz
- Points/Dec: constant calculation point in 1 decade, here 10 points.

The setting method for geometric information of lines is the same as the Bergeron model.

An example of calculated line constants is shown next. The format of line constants is different from line constants (Figure 2.20).

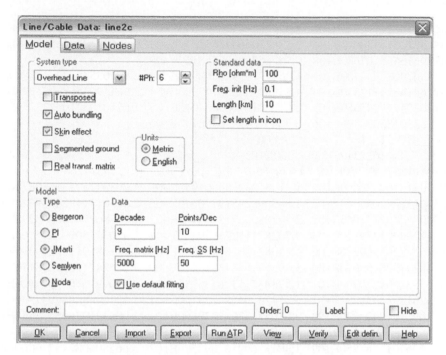

Figure 2.20 J. Marti setting in ATPDraw.

```
 −1IN___AOUT__A              2.  0.00                         −2 6
     24          7.6398599099200032000E+02
 1.75644935255282330E+04  2.13770954194475930E+04 −3.64488176214450650E+04
 4.94864953079189220E+01  6.20881202228397110E+02  2.61347936239796950E+03
 1.06446596255732570E+04  5.56483351718696580E+03  9.61544227858097470E+03
 1.64837914474179050E+04  5.96654317870005400E+04  4.84716733816883880E+05
 1.04609247552440010E+06  2.06603388986925640E+06  2.70144624683398050E+06
 5.28779897867655380E+06  2.03830524591382100E+06  2.86059675266138650E+06
 4.67738978494544610E+06  5.75897667709425560E+06  9.63263244143215570E+06
 1.72944612796751150E+07  3.91134948735416310E+07  2.30868878964187410E+08
 1.41932907125968020E+00  1.56999988850242270E+00  1.53443313993469150E+00
 2.42506222838704400E+00  6.04343530774100040E+00  2.08840784254574070E+01
 6.56597595574143900E+01  8.21531555331479950E+01  1.35214190582626970E+02
 7.53913977257779040E+02  2.70493312768724900E+03  1.13604856565846690E+04
 5.02950884740586040E+04  1.04812225198132360E+05  2.75339949282041750E+05
 5.49495598081170810E+05  8.61025381440750320E+05  1.21232041836172110E+06
 1.93177936890560920E+06  2.48934709176647660E+06  4.06343682673577450E+06
 7.36530641369509510E+06  1.66209975471609050E+07  9.83908682559759170E+07
     11          3.3452065056893213000E−05
 7.09045971805790030E+00  8.29013079087817460E+01  2.05218646853829370E+02
```

8.27684059224196970E+02   2.02596286316561600E+03   1.71214370952625320E+04
4.23815272516980040E+04   2.53827066343654530E+05   4.49644376013503640E+05
9.74979676558507680E+08 −9.75745799822558160E+08
1.25483679032792700E+03   1.36977105990960950E+04   2.89403535922992780E+04
2.74776414529713330E+04   4.55959309852172080E+04   2.96419035686776390E+05
1.96146205717745560E+05   7.61154199065004010E+05   1.67523034109223260E+06
6.75915330185273570E+06   6.76591245515458190E+06
−2IN___BOUT__B   2. 0.00   −2 6
      15   3.9456222106786458000E+02
2.99204109563495880E+03   1.47468507580421000E+02 −1.92945146689272810E+03
7.80440714107136840E+02   1.31985349384001050E+03   6.79337259131680870E+02
5.68146138742036440E+02   1.86309829519770940E+02   2.39301214924039020E+02
2.94320345497146040E+02   2.70788995218796460E+02   1.37786074042563290E+02
6.09512239981379120E+03   3.70287113225839350E+04   1.42035350097225560E+06
1.25816147265869250E+00   1.94683183758184030E+00   1.72866601165299990E+00
1.63048293326194060E+00   5.39554164013315510E+00   7.84688149832078170E+00
1.14863334604270960E+01   1.59848711726535930E+01   2.32099101677073420E+01
3.86162618954434790E+01   6.21115063584223750E+01   1.12360880937937340E+02
4.27120951912425330E+03   2.61053133008279660E+04   1.00432192695335540E+06
       9   3.3353569212064511000E−05
2.47347067017885850E+02   1.40133699531098550E+03   5.66580603243716000E+03
5.09644619418256840E+04   9.19201196234028320E+04   8.70850634276895320E+05
4.92415836815076140E+06   2.32285736449953130E+07   1.10092047367656440E+08
3.97983369524935620E+04   2.22313709917902540E+05   4.52916300172152410E+05
2.42236089878577180E+06   1.75381055169851870E+06   8.96838074267333750E+06
2.77902908309910110E+07   8.60795688641980440E+07   3.08138594955910330E+08
−3IN___COUT__C   2. 0.00   −2 6
      20   3.7816769050640863000E+02
2.76777910671226800E+03   3.19970733700880200E+02 −4.48851933075702350E+03
3.38579513491869560E+03   8.86525969650747810E+02 −5.58844802144627370E+03
6.92710802790131490E+03   1.83212705278188420E+02   4.83783036871584780E+02
4.25714960801934070E+02   5.86157177598852060E+02   5.18503321319391600E+02
9.81902029327581400E+01   2.40718591658280730E+02   3.14021678020258780E+02
2.70325573454710270E+03   3.18519377629979140E+03   5.85928462694478460E+03
2.35245344588584700E+04   4.05493776380774620E+05
1.25992761407011480E+00   1.96063297159097490E+00   1.75123520042841710E+00
1.69488563149130700E+00   4.94664773364731490E+00   9.55007218309695990E+00
9.34879784362447990E+00   1.34580881251905760E+01   1.98913142485725450E+01
3.14577870495336140E+01   6.38810246288385240E+01   1.10035910301716270E+02
1.60161390036292150E+02   3.65930495760880380E+02   4.75594017789749500E+02
3.88748749091944730E+03   4.64665428838566280E+03   8.53804721738672010E+03
3.42713404536987730E+04   5.91203537001200250E+05
       9   −3.3357879651165086000E−05
−2.21998133608119420E+02 −2.27342777649989560E+03 −3.20236258741104300E+03
−2.32219195762859560E+04 −5.18998064583304000E+04 −3.99895274544087820E+05

```
−4.30777098695122260E+06 −2.65697164933130930E+07  9.12715176732486630E+00
 3.80203870724501680E+04  3.96296040781158080E+05  5.40228585390166960E+05
 3.92477868761868770E+06  2.23021404315216330E+06  8.17104251386753000E+06
 5.30253207385864260E+07  9.63865815985234380E+07  1.98691765316513630E+13
−4IN___DOUT__D            2.  0.00          −2 6
     19    3.3216156753891221000E+02
 3.32679816953891260E+03    1.36640206420175260E+03 −4.30389100804873030E+03
 1.46179736037509060E+03    2.53726725080325880E+03 −5.71788781355530770E+02
 8.50012783054611650E+02    7.64777577760723660E+01  2.30117415905815930E+02
 3.06333251678185720E+02    2.73258447411545770E+02  6.49240242920953390E+01
 1.37486725868792400E+02    1.66562225251179370E+02  2.22772272047848820E+02
 5.65528853779869500E+03    4.47803068414749120E+03  4.95493115104619850E+03
 1.32034477990225480E+05
 1.31455767119944620E+00    1.88884200340097050E+00  1.78111974528292730E+00
 1.71644352333832040E+00    6.76166440256223300E+00  8.11335741202654330E+00
 1.35662187646851890E+01    1.79711915741429420E+01  2.63965646326654130E+01
 4.47244203787357510E+01    7.11044086568985990E+01  1.17505321023316410E+02
 2.35369295931236420E+02    2.82363426458155400E+02  3.77068241870125010E+02
 9.09411805948983060E+03    6.91110684770364700E+03  7.83023215395506030E+03
 2.08846067652582070E+05
      9    −3.3356929796641675000E−05
−1.87012935699362940E+02  −3.14461868705284180E+03  −4.95779795674018440E+03
−5.69753025757968060E+04  −1.09861613450616440E+05  −1.14003399342954690E+06
−1.15392962860063780E+07 −8.61885459056655910E+07  9.02489130826209700E+00
 3.29820351724520370E+04   5.59480720870455260E+05  8.64150908942334470E+05
 9.86335264955757000E+06   4.82149845230978080E+06  2.42116055759465960E+07
 1.23636239563861460E+08   3.44131811882111250E+08  6.28318530721700470E+13
−5IN___EOUT__E            2.  0.00          −2 6
     19    2.9471645239150092000E+02
 3.04608646375060740E+03    3.85075805560429790E+02  1.23201563500287260E+03
−2.88301172023220400E+03    2.58413315972866180E+03 −5.26048990242245850E+02
 6.34895119580601660E+02    2.61420872948880630E+02  2.14686896154274000E+02
 3.01395001442612620E+02    2.75441714106578900E+02  6.10376245361192390E+01
 1.33906276872624150E+02    1.69396715781134620E+02  1.43308768621841800E+03
 2.99087497757312530E+03    4.17252559627266240E+03  9.60099715603003820E+03
 1.28488490897544430E+05
 1.31939418222998110E+00    1.99910596270501690E+00  1.73781074980951680E+00
 1.79955404194475490E+00    7.32315915670604150E+00  8.89804088246597050E+00
 1.41877831687642200E+01    2.08887076512220280E+01  2.91054594535489310E+01
 4.89722998616522320E+01    7.93752529546837590E+01  1.26834163588896130E+02
 2.50290242275297030E+02    3.16852218220506470E+02  2.49685392932030210E+03
 5.19435278129407020E+03    7.35358894458953910E+03  1.68605988481572090E+04
 2.25345768446158060E+05
     13    3.3345738274281065000E−05
 2.59340140778726460E+01    3.37369352604558400E+02  6.60788166962460880E+02
 2.41181053190797320E+03    3.93704194589121790E+03  5.27579832290847480E+03
```

3.38744718130088860E+03   6.29203906081054810E+03   1.26473396260911370E+05
1.35591748912418820E+06   8.43621652354194970E+06   2.42607264330695090E+07
8.67013469343307140E+08
8.74759146206365220E+03   1.13606116013118360E+05   2.24287656773600550E+05
8.02729820233730250E+05   1.34530709863136500E+06   1.78751838160319040E+06
1.13700793302550420E+06   2.14072102775288420E+06   5.42768468281883000E+06
2.86352364743568820E+07   9.23733307088016720E+07   2.67363604461298230E+08
1.19838585714910410E+09
–6IN___FOUT__F   2. 0.00   –2 6
   20   3.00817646763869850000E+02
3.70857556087103790E+03   8.86985299082655270E+01–1.39740213064266640E+03
–6.07581501853979490E+02   1.88556379139355110E+03   7.42311569945918220E+02
–2.80035191411265940E+02   2.91145939383249130E+02   3.45992630139773890E+02
2.19371129077163290E+02   2.97758634059481320E+02   2.73043679523143280E+02
6.27327368361363470E+01   1.34348467620353800E+02   1.61559436548511140E+02
2.41079010411799000E+02   7.49718302379666150E+03   5.80760793232300690E+03
3.51096220670131560E+03   1.29404478005025380E+05
1.34416280851345200E+00   2.03184903232078760E+00   1.66356650222744640E+00
1.78593295185650880E+00   6.97909423386909290E+00   9.86597593114704630E+00
1.17348936839463980E+01   1.61248179208375080E+01   2.13897014776734760E+01
2.92276510460420730E+01   4.81543820985171680E+01   7.74514977923254410E+01
1.25463750599228560E+02   2.44787530925921200E+02   2.98641570128496140E+02
4.36797244889617330E+02   1.29547954100466870E+04   9.91187765802387730E+03
5.99342270014167840E+03   2.22566358376151240E+05
   13   3.33458148102308360000E–05
2.54970001520574100E+01   3.26184910170085120E+02   6.49105227487062280E+02
2.12639146520084110E+03   4.07594926475122250E+03   5.31370033851396690E+03
5.75105757361272030E+03   4.35935195503303840E+03   1.24724423784763490E+05
1.30903102511541960E+06   8.13029298079738300E+06   2.46821016424198230E+07
8.72012717367038850E+08
8.74652864236458480E+03   1.17731870268556970E+05   2.24760716518480850E+05
7.22434341161434770E+05   1.35004215090710760E+06   1.82902683638393440E+06
2.06816129362340220E+06   1.49143889475726920E+06   5.44066062072633490E+06
2.81315933761330170E+07   9.07277028467590660E+07   2.72802134855029940E+08
1.19938686378418250E+09
 0.25956082 –0.23608915   0.35367333 –0.37706278 –0.35599675 –0.35880969
–0.01942457   0.00092063 –0.35407376 –0.40740363   0.00782495   0.00664004
 0.32871687 –0.31450666   0.09427163 –0.15022206   0.57415285   0.56545558
–0.00917400 –0.03328584 –0.11166193 –0.14007565 –0.01631131   0.00595174
 0.56930828 –0.58671857 –0.32038975   0.26209366 –0.20709645 –0.22653698
 0.00000000   0.00000000   0.35431883   0.28454350 –0.01996485 –0.01082001
 0.25956082   0.23608915   0.35367333   0.37706278   0.35599675 –0.35880969
–0.01942457 –0.00092063 –0.35407376   0.40740363 –0.00782495   0.00664004
 0.32871687   0.31450666   0.09427163   0.15022206 –0.57415285   0.56545558
–0.00917400   0.03328584 –0.11166193   0.14007565   0.01631131   0.00595174
 0.56930828   0.58671857 –0.32038975 –0.26209366   0.20709645 –0.22653698
 0.00000000   0.00000000   0.35431883 –0.28454350   0.01996485 –0.01082001

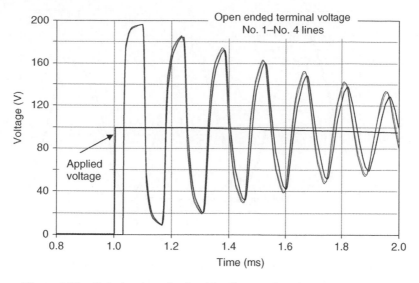

Figure 2.21 Calculated result of making lines, various frequency modeling.

Figure 2.21 shows the voltage waveforms making the source to the No. 1–No. X lines listed below. The J. Marti model is suitable for calculating high-frequency phenomena.

No. 1	line	fm = 50 Hz	fs = 50 Hz	fl = 10 Hz
No. 2	line	500 Hz	50 Hz	10 Hz
No. 3	line	5000 Hz	50 Hz	10 Hz
No. 4	line	50,000 Hz	50 Hz	10 Hz
No. X	line	1 MHz	100 kHz	0.1 Hz

(for a wide range up to 10 MHz)

Next, Figure 2.22 shows the calculated short-circuit currents. We can see some differences between the J. Marti model and the line constants model.

Figure 2.23 shows the DC voltage at small capacitive current interruption for no load transmission line. The DC voltage decays. It is necessary to be careful when calculating the reclosing of the no-load transmission line.

2.1.2   Underground Cables—Cable Parameters

2.1.2.1   Pi Equivalents Model

The cable can be represented by a pi equivalent model, the same as the overhead transmission line in case of a grounded cable sheath.

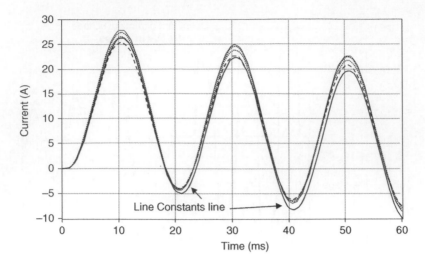

Figure 2.22 Calculated result of short-circuit current at J. Marti and line constants.

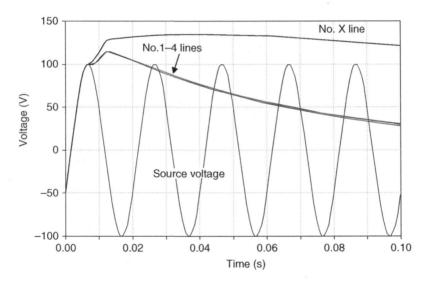

Figure 2.23 DC voltage at small capacitive current interruption for no load with J. Marti lines.

Inductance and capacitance of the cable are shown in Equation (2.26).

$$\text{Inductance } L, \qquad\qquad \text{Capacitance } C$$

$$L = \frac{\mu_0 \mu_s}{2\pi} \ln \frac{r_2}{r_1} (\text{H}/\text{m}), \quad C = \frac{2\pi\varepsilon_0 \varepsilon_s}{\ln \dfrac{r_2}{r_1}} \tag{2.26}$$

μ_0: vacuum permeability$=4\pi \times 10^{-7}$ (H/m)
μ_s: relative permeability of insulator $= 1.0$
ε_0: vacuum dielectric constant$=1/36\pi \times 10^{-9}$ (F/m)
ε_s: relative dielectric constant of insulator
r1: outer radius of conductor (mm)
r2: inner radius of sheath (mm).

The relative permittivity of insulators is as follows (XLPE = (cross-linked polyethylene): Oil-filled paper: 3.7, XLPE: 2.3

2.1.2.2 Distortionless Line Modeling

Surge impedance, Propagation velocity

$$Z = \sqrt{\frac{L}{C}}\,(\Omega) \qquad v = \frac{1}{\sqrt{LC}}\,(\mathrm{m}/\mathrm{s}) \tag{2.27}$$

The calculation method of inductance and capacitance for XLPE cable is different from that of oil-filled (OF) cable.

2.1.2.2.1 Case of OF Cable

In OF cable, the screen is very thin, so we can neglect this.
The inductance and capacitance will be expressed in Equation (2.28).

Inductance L \qquad\qquad Capacitance C

$$L = \frac{\mu_0 \mu_s}{2\pi} \ln \frac{r_2}{r_1}\,(\mathrm{H}/\mathrm{m}), \qquad C = \frac{2\pi\varepsilon_0\varepsilon_s}{\ln \dfrac{r_2}{r_1}}\,(\mathrm{F}/\mathrm{m}) \tag{2.28}$$

$$Z = \sqrt{\frac{L}{C}} = \frac{60}{\sqrt{\varepsilon_s}} \ln \frac{r_2}{r_1}\,(\Omega), \qquad \upsilon = \frac{1}{\sqrt{LC}} = \frac{300}{\sqrt{\varepsilon_s}}\,(\mathrm{m}/\mu\mathrm{s}) \tag{2.29}$$

2.1.2.2.2 Case of XLPE (Cross-Linked Polyethylene) Cable

In XLPE cable, the thickness of the conductive tape cannot be neglected.
It is necessary to treat the conductive tape as a conductor in the calculation of capacitance of the cable. Otherwise, it is necessary to treat it as an insulator when calculating inductance.
The inductance and capacitance will be expressed as Equations (2.30) and (2.31). These equations are different from the case of OF cable.

$$L = \frac{\mu_0 \mu_s}{2\pi} \ln \frac{r_2}{r_1}\,(\mathrm{H}/\mathrm{m}) \tag{2.30}$$

$$C = \frac{2\pi\varepsilon_0\varepsilon_s}{\ln\dfrac{r_o}{r_i}}\left(\mathrm{F\,/\,m}\right) \tag{2.31}$$

$$Z = \sqrt{\frac{L}{C}} = \frac{60}{\sqrt{\varepsilon_s}}\sqrt{\ln\frac{r_o}{r_i}\times\ln\frac{r_2}{r_1}}\left(\Omega\right) \tag{2.32}$$

$$\upsilon = \frac{1}{\sqrt{LC}} = \frac{300}{\sqrt{\varepsilon_s}}\sqrt{\ln\frac{r_o}{r_i}\,/\,\ln\frac{r_2}{r_1}}\left(\mathrm{m\,/\,\mu s}\right), \tag{2.33}$$

where r_1 = outer radius of conductor (mm), r_2 = inner radius of sheath (mm), r_i = outer radius of inner semiconductive layer (mm), and r_o = inner radius of outer semiconductive layer (mm).

2.1.2.3 Cable Constants

Figure 2.24 shows an example of the structure of 220 kV XLPE, and Table 2.2 gives an example of the cable data.

It is necessary to input the data for calculating these cable constants using the cable constants subroutine (Figure 2.25 and Table 2.3).

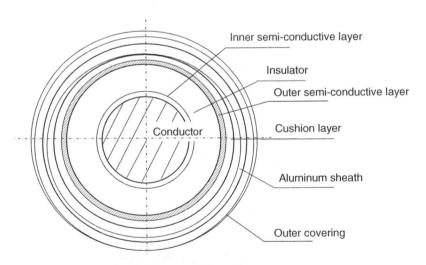

Figure 2.24   Example of cable structure.

Table 2.2 Example of cable data.

Number of core		—	1
Conductor	Cross-sectional area	mm²	600
	Outer radius	mm	29.5
Thickness of inner semiconductive layer		mm	1.0
Thickness of insulator		mm	27.0
Thickness of outer semiconductive layer		mm	1.0
Thickness of cushion layer		mm	3.0
Thickness of aluminum sheath		mm	2.5
Thickness of outer covering		mm	5.0
Core DC resistance		Ω/km	0.0308
Capacitance		μF/km	0.15

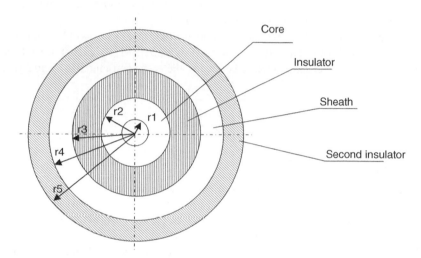

Figure 2.25 Input data of cable in EMTP.

Table 2.3 Example input data of cable.

	Radius	(mm)	
Inner radius of core	r1	0	—
Outer radius of core	r2	14.75	Radius of conductor 14.75 mm = 29.5/2 mm
Radius of insulator	r3	46.75	Radius of conductor 14.75 mm + thickness of inner semiconductive layer 1 mm + thickness of insulator 27 mm + thickness of outer semiconductive layer 1 mm + thickness of cushion layer 3 mm = 46.75 mm
Inner radius of sheath	r4	49.25	r3 + aluminum sheath 2.5 mm = 49.25 mm
Radius of second insulator	r5	54.25	r4 + outer covering 5 mm = 54.25 mm

Conductor:	resistivity	$1.848\text{E}{-8}$ $(\Omega\cdot\text{m})$
		(DC resistivity $0.0308\,\Omega/\text{km}\times600\,\text{mm}^2$)
	relative permeability	1.0
Insulator:	relative permeability	1.0
	relative dielectric constant	2.66 $\left(=\dfrac{\ln\left(46.75\,/\,14.75\right)}{\ln\left(42.75\,/\,15.75\right)}\times2.3\right)$
Sheath:	resistivity	$2.83\text{E}{-8}$ $(\Omega\cdot\text{m})$
		(Resistivity of Al $2.83\,\mu\Omega\cdot\text{cm}$)
	relative permeability	1.0
Outer covering:	relative permeability	1.0
	relative dielectric constant	4.0

In the case of OF cable, the screen is constructed with carbon paper or metallic tape. The user can neglect the screen due to its thinness. The user can treat the outer radius of the insulator as equal to the inner radius of the sheath. Also, the user can treat the inner radius of the insulator as equal to the outer radius of the conductor.

However, in the case of XLPE cable, it is necessary to count the thickness of both inner and outer semi-conductive. The Electromagnetic Transients Program (EMTP) cannot treat these semi-conductive layers, so the user must convert the relative dielectric constant using Equation (2.34),

$$\varepsilon_t = \frac{\ln\left(\dfrac{r_3}{r_2}\right)}{\ln\left(\dfrac{r_{so}}{r_{si}}\right)}\times\varepsilon_s, \tag{2.34}$$

where r_{si} = inner radius of insulator (radius of conductor + thickness of inner semiconductive layer, and r_{so} = outer radius of insulator.

The following figures show examples of ATPDraw input. Figures 2.26 and 2.27 show a single cable Bergeron model with the sheath grounded.

Figures 2.28 and 2.29 show a single-cable Bergeron model with the sheath treated as conductor.

EMTP-ATP input data:

```
BEGIN NEW DATA CASE
CABLE CONSTANTS
BRANCH  IN___AOUT__AIN___BOUT__B
  2    1     1        1              2  0  0
  2
      0.0   0.01475   0.04675   0.04925   0.05425
  1.848E–8    1.        1.       2.66     2.83E–8    1.    1.    4.
     10.        0.0
       100.             50.              100.     2
BLANK CARD ENDING FREQUENCY CARDS
$PUNCH
BLANK CARD ENDING CABLE CONSTANTS
BEGIN NEW DATA CASE
BLANK CARD
```

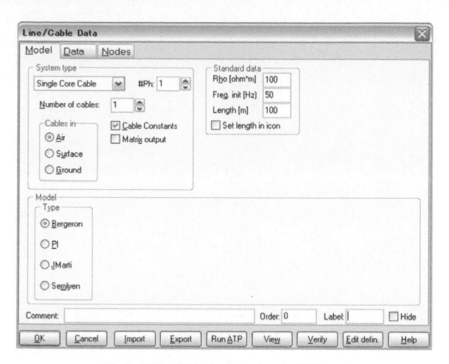

Figure 2.26   Cable data input form in ATPDraw.

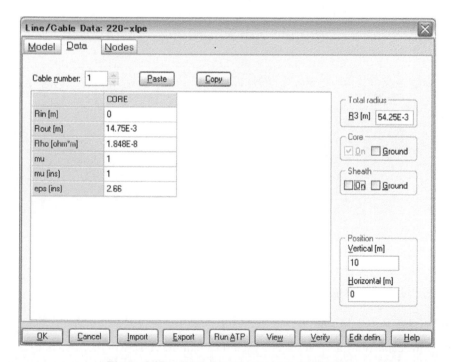

Figure 2.27   Cable data input form in ATPDraw.

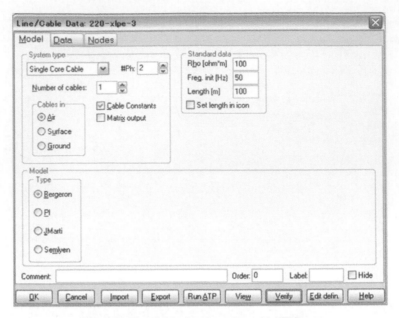

Figure 2.28   Cable data input form in ATPDraw.

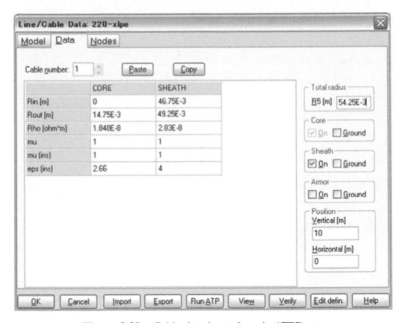

Figure 2.29   Cable data input form in ATPDraw.

2.1.2.4   Cross-Bonded Cable

In a long-line cable transmission system, each sheath of the cable will be crossed, as shown in Figure 2.30. EMTP-ATP has this cross-bonded cable model but ATPDraw does not. The user needs to make that model manually.

The user can cross the sheath manually using "Transpose" in ATPDraw (Figures 2.31 and 2.32).

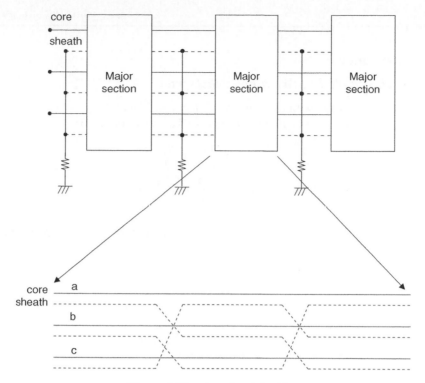

Figure 2.30 Cross-bonded cable.

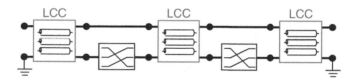

Figure 2.31 Cross the sheath using "Transpose" in ATPDraw.

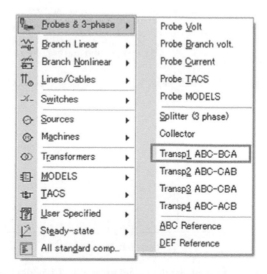

Figure 2.32 Transpose in the ATPDraw menu.

2.2   Transformer

2.2.1   Single-Phase Two-Winding Transformer

A single-phase transformer with two-winding is shown in Figure 2.33.
   In Figure 2.33, voltages and currents are expressed as Equation (2.35).

$$v_1 = i_1 r_1 + n_1 \frac{d\phi_1}{dt} + n_1 \frac{d\phi_0}{dt}$$

$$v_2 = i_2 r_2 + n_2 \frac{d\phi_2}{dt} + n_2 \frac{d\phi_0}{dt} \tag{2.35}$$

Here, as we can set $n_1\phi_1 = l_1 i_1$, $n_1\phi_0 = L_0 i_0$ $n_2\phi_2 = l_2 i_2$, Figure 2.33 will be converted to Figure 2.34.

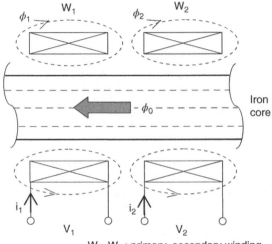

W_1, W_2 : primary, secondary winding

Φ_1, Φ_2 : fluxes crossed only W_1, W_2

Φ_0 : common flux

v_1, v_2, i_1, i_2: voltage and current of winding

Figure 2.33   Single phase with two-winding transformer.

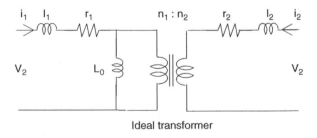

Ideal transformer

Figure 2.34   Equivalent circuit of single-phase two-winding transformer.

2.2.1.1 Transformer Model

EMTP has the "Transformer model" shown in Figure 2.35. The leakage inductances (L1, L2) will be calculated from percentage impedance, %Z.

2.2.1.2 ATPDraw Input

Figure 2.36 shows a transformer menu in ATPDraw.

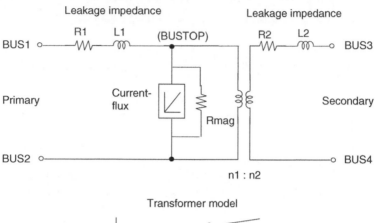

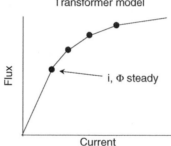

Figure 2.35   EMTP transformer model.

Figure 2.36   Transformer menu in ATPDraw.

Figure 2.37 presents a saturable one-phase model.
Figures 2.38 and 2.39 give examples of current-flux characteristics.

2.2.1.3   Case of Hysteresis Modeling

In the case of modeling hysteresis characteristic of the core, users should connect the Type 96 pseudo-nonlinear hysteretic inductor to the outside of the transformer model (Figure 2.40). The primary leakage impedance should be set at a small value.

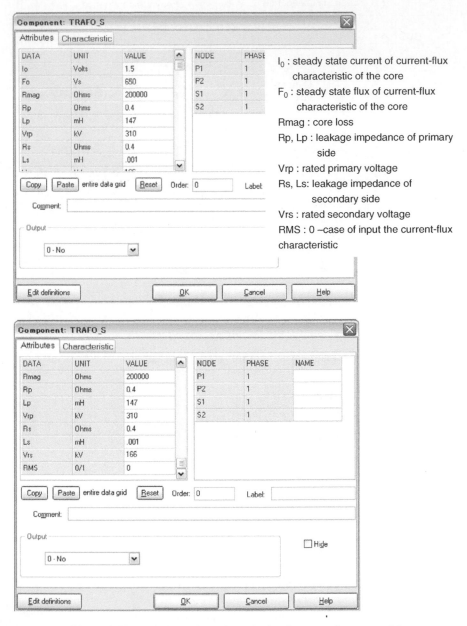

Figure 2.37   Data input form for a single-phase transformer model.

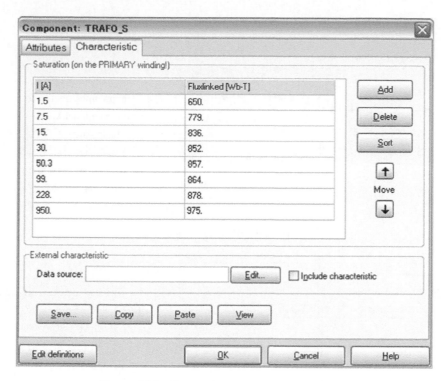

Figure 2.38 Current flux for iron core input form.

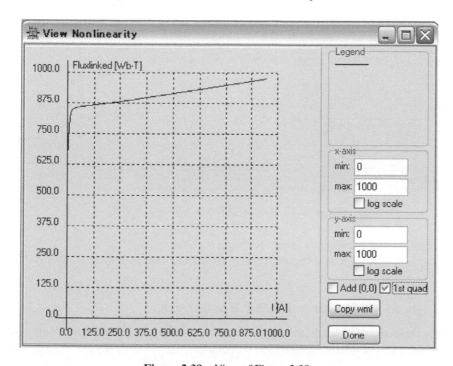

Figure 2.39 View of Figure 2.38.

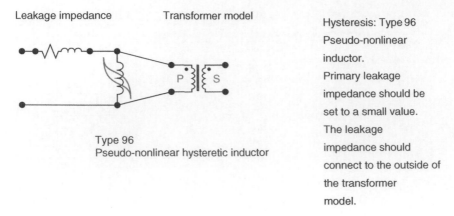

Leakage impedance   Transformer model

Hysteresis: Type 96
Pseudo-nonlinear
inductor.
Primary leakage
impedance should be
set to a small value.
The leakage
impedance should
connect to the outside of
the transformer
model.

Type 96
Pseudo-nonlinear hysteretic inductor

Figure 2.40   Hysteresis modeling.

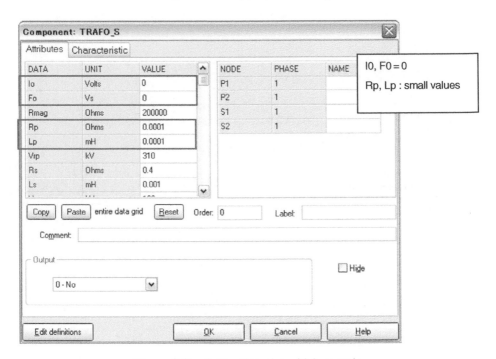

Figure 2.41   Setting the value with hysteresis.

The setting of the transformer model is shown in Figure 2.41.
Figures 2.42, 2.43, and 2.44 show a Type 96 pseudo-nonlinear inductor.

2.2.2   Single-Phase Three-Winding Transformer

The equivalent circuit of single transformer with three-winding is shown in Figure 2.45. For example, we try to calculate each leakage inductance.

Percentage impedances:

HV-MV	21.3% (at 300 MVA base)
HV-LV	15.0% (at 90 MVA base)
MV-LV	13.1% (at 90 MVA base)

$$Z_{HM} = 0.213 \times \frac{275\,\mathrm{kV}^2}{300\,\mathrm{MVA}} = 53.7\,\Omega$$

$$Z_{HL} = 0.150 \times \frac{300\,\mathrm{MVA}}{90\,\mathrm{MVA}} \times \frac{275\,\mathrm{kV}^2}{300\,\mathrm{MVA}} = 126\,\Omega \qquad (2.36)$$

$$Z_{ML} = 0.131 \times \frac{300\,\mathrm{MVA}}{90\,\mathrm{MVA}} \times \frac{275\,\mathrm{kV}^2}{300\,\mathrm{MVA}} = 110\,\Omega$$

$$\begin{array}{ccc} Z_1 + Z_2 = 53.7 & & Z_1 = 34.8 \\ \rightarrow \quad Z_1 + Z_3 = 126\,\Omega \rightarrow & Z_2 = 18.9\,\Omega \\ Z_2 + Z_3 = 110 & & Z_3 = 91.2 \end{array} \qquad (2.37)$$

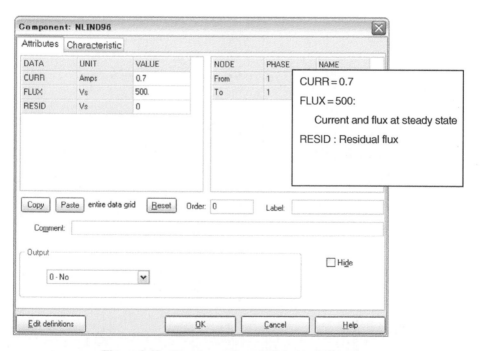

Figure 2.42   Pseudo-nonlinear inductance data setting.

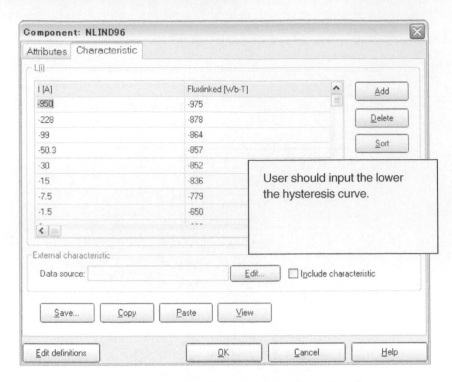

Figure 2.43 Pseudo-nonlinear inductance flux data setting.

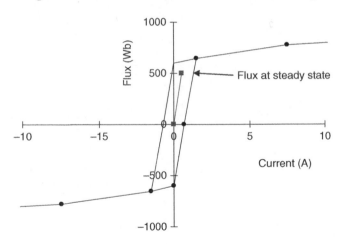

Figure 2.44 Example of hysteresis.

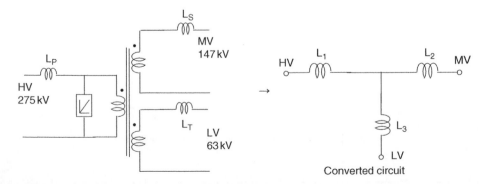

Figure 2.45 Equivalent circuit for a three-phase transformer.

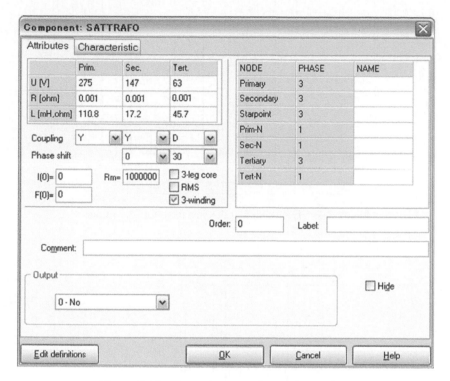

Figure 2.46   Three-phase transformer data setting.

$$L_1 = \frac{Z_1}{\omega} = 110.8\,\text{mH}$$

$$L_2 = \frac{Z_2}{\omega} \times \left(\frac{147\,kV\,/\,\sqrt{3}}{275\,kV\,/\,\sqrt{3}}\right)^2 - 17.2\,\text{mII} \tag{2.38}$$

$$L_3 = \frac{Z_3}{\omega} \times \left(\frac{63\,kV}{275\,kV\,/\,\sqrt{3}}\right)^2 = 45.7\,\text{mH}$$

Figure 2.46 shows a saturable three-phase model.

2.2.3   Three-Phase One-Core Transformer—Three Legs or Five Legs

In a three-phase and three-leg transformer, some part of the zero-sequence flux does not flow to the core, as shown in Figure 2.47. In this case, zero-sequence current-flux characteristics will not be saturable and will be linear.

This equivalent circuit is shown in the *ATP-Rule Book* (Figure 2.48).

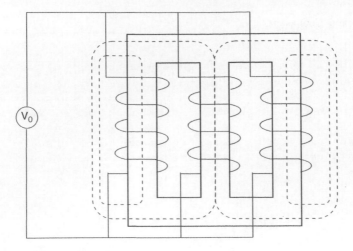

Figure 2.47 Explanation of flux in a three-leg transformer.

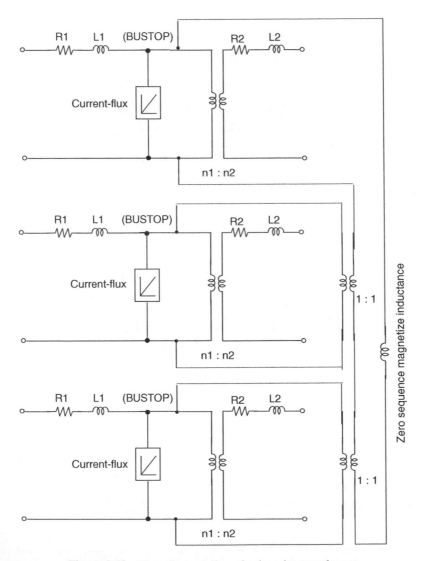

Figure 2.48 Example modeling of a three-leg transformer.

2.2.4   Frequency and Transformer Modeling

Transient phenomena occurring in power systems have frequency ranges from DC to several 10 MHz. We can make a universal transformer model covering the wide frequency range, but this is not realistic.

It is known that the transformer model is suitable for use in a frequency range from AC to several kilohertz.

In the high-frequency range, we will not neglect the effect of stray capacitances. In this case, the user should model the same capacitances as shown in Figure 2.49.

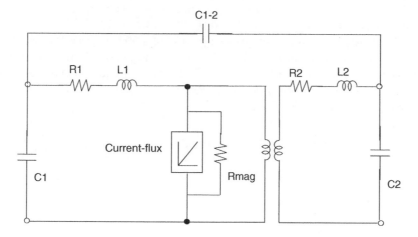

Figure 2.49   Transformer model with capacitances.

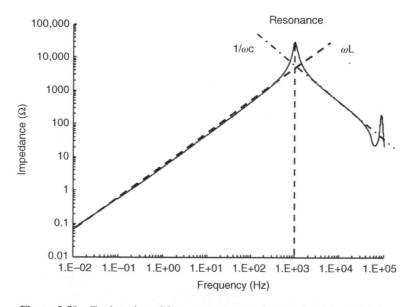

Figure 2.50   Explanation of frequency response for transformer impedance.

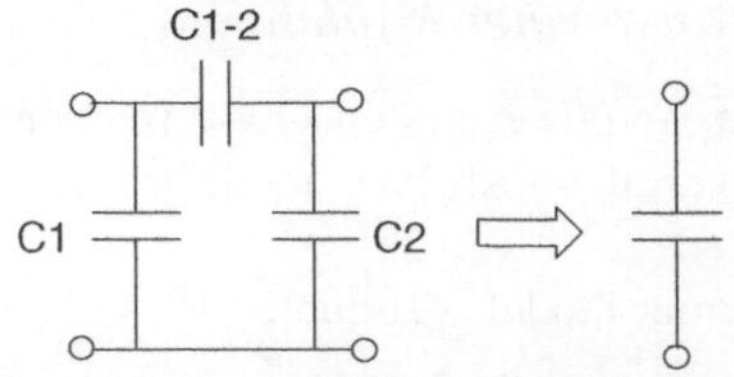

Figure 2.51 Transformer model at quite high frequency.

In the higher frequency ranges, such as in lightning surges, we can neglect core inductance and leakage inductance.

Figure 2.50 shows the measured frequency response of a 50 MVA transformer. We can see the resonance frequency is about 1 kHz. In the lower frequency under the resonance frequency, the impedance is nearly equal to ωL. Otherwise, in the upper frequency above the resonance frequency, the impedance is nearly equal to $1/\omega c$. We can model this transformer as only capacitance over several 100 kHz (Figure 2.51).

3

Transient Currents in Power Systems

3.1 Short-Circuit Currents

Short-circuit currents in power systems concern engineering work on circuit breaker switching, relay protection, electromagnetic force withstanding, and so on. In most cases of such works, the effect of the high-frequency current component created by charging/discharging of the capacitance in the circuit can generally be disregarded; therefore, the relevant circuits can be modeled by inductance(s) and resistor(s) only. Figure 3.1 shows the most simplified circuit model. Regarding the circuit, the following analytical solution, Equation (3.1), is easily obtained.

$$i = I\left\{\sin(\omega t + \alpha - \theta) - \exp\left(-R/L\right)t \cdot \sin(\alpha - \theta)\right\},$$

$$\text{where} \quad \theta = \tan^{-1}\left(\omega L/R\right)$$

$$\sin(\alpha - \theta): DC\ component\ at\ t = 0\ current\ starting$$

$$\omega: AC\ source\ angular\ velocity$$

(3.1)

Depending on each point of source voltage wave timing of short-circuiting, the short-circuit current has a decaying DC component with a maximum crest value of that of the power frequency component. In actual power systems, due to some circuits having individual L/R values (DC decaying time constants) connected in parallel, current waveforms may differ a little from Equation (3.1); nevertheless, the circuit as shown in Figure 3.1 is mostly applicable with sufficient practical accuracy.

Power System Transient Analysis: Theory and Practice using Simulation Programs (ATP-EMTP), First Edition.
Eiichi Haginomori, Tadashi Koshiduka, Junichi Arai, and Hisatochi Ikeda.
© 2016 John Wiley & Sons, Ltd. Published 2016 by John Wiley & Sons, Ltd.
Companion website: www.wiley.com/go/haginomori_Ikeda/power

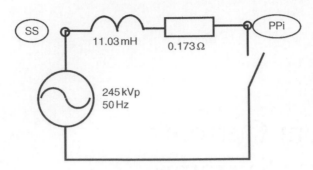

Figure 3.1   Simplified AC circuit.[1]

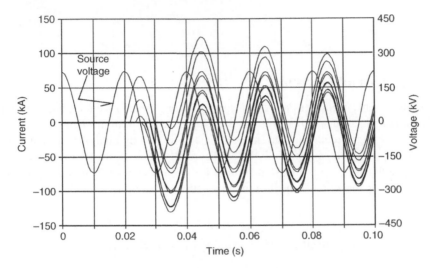

Figure 3.2   Short-circuit currents in the circuit in Figure 3.1.[1]

Applying ATP-EMTP, calculated examples are shown in Figure 3.2 (see also the EMTP data file[1]). As shown in the figure, each current starts from a zero value due to the continuity principle of current in inductance, while a constant AC component value, irrespective of the DC component value and short-circuiting timing, is involved in each current waveform.

In practical system circuits, where the circuits are mostly of three phases, both positive/ negative and zero sequence parameters are to be considered. Also for practical systems, discharging currents from parallel capacitances, such as transmission lines, cables, or shunt capacitor banks, may not be, and in limited special cases, even insignificant.

Such discharging currents have components of several hundred hertz and most decay after several tens of microseconds from the instance of short-circuit. But in special cases, they may not be negligible for such times when currents are interrupted by circuit breakers. A typical example, corresponding to an extremely high power density network near a megalopolis, is shown in Figure 3.3a (circuit) and 3.3b (calculated currents), where an EHV (extra high voltage) substation bus bar is short-circuited in three phases [1].[2] In the circuit, extremely high capacitances, such as EHV cables via a certain length of overhead line (20 km) and high-capacity shunt capacitor banks in the tertiary winding side of the transformer, are connected. Care should

[1] ATPData3-01.dat, DATA3-01.acp: Short-circuit current calculation in a simplified power frequency source circuit.
[2] ATPData3-02.dat, DAT3-02.acp: High-capacity EHV substation bus bar short-circuit transient current.

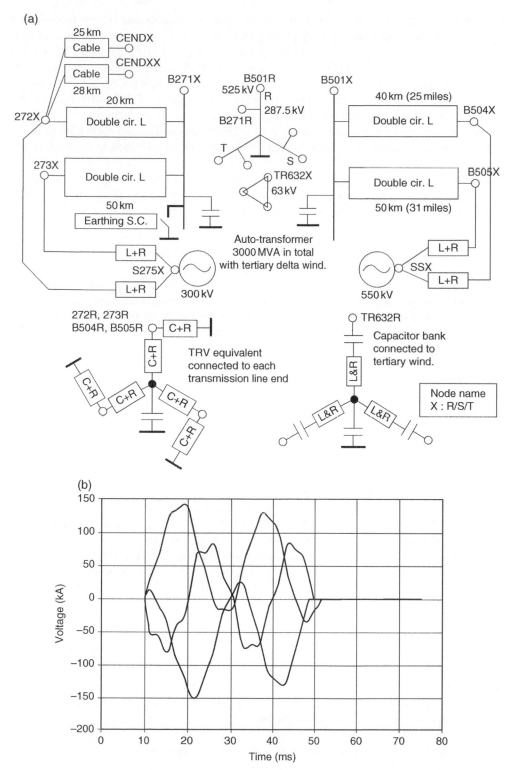

Figure 3.3 EHV high-power/density network and short-circuit currents [1].[2] (a) EHV high-power/density system network. (b) Short-circuit currents in such a network.

be taken in such calculations with regard to the damping of the transient current frequency of capacitance discharging current. The frequency of the transient is in the order of several hundred hertz, so the losses in transformers, transmission lines, cables, and so on, are to be based on that frequency range. The calculation is made for a 550/300 kV substation, the capacity of its transformer is (in total) 3 GVA, and where a total of 600 MVA of capacitor bank is connected to the tertiary side of the transformer. A in total of 53 km (25 + 28 km), of EHV cable is connected to the 300 kV bus bar via 20 km of overhead transmission line. In the case of damping resistances in the system, circuits were carefully adjusted based on the transient current frequencies.

Applying the "Fourier On" menu, GTPPLOT or Plot XY, the Fourier spectrum is easily obtained and the wave shape in Figure 3.3b (with maximum transient component) has 10% of fifth harmonics, which can yield 50% of the enhancement of the *di/dt* value at the current interruption instant by a circuit breaker, which may significantly affect a certain type of circuit breaker's performance (For details of the system parameters applied, see the data file[2]).

Note:

Such short-circuit current distortion is significant where a very high capacitance exists via a certain value of inductance, for example, the transmission line, transformer, series reactor of shunt capacitor bank, and so on. Also, it should be noted that in any other case than three-phase short-circuiting, the damping of the high-frequency component is very quick due to the zero-sequence circuit parameters.

3.2   Transformer Inrush Magnetizing Current

The transformer's ion core is designed so as not to be saturated under normal (steady state) service conditions. But on special occasions, for example, at switching on to the system, the magnetic flux passes into the saturated region; therefore, very high current flows through, which is called the inrush magnetizing current. Such a current is often explained by applying the diagram as shown in Figure 3.4. For EHV high-capacity transformers, the current values often record up to several thousand A, while the steady state magnetizing currents are even lower than 1 A. The calculation applying ATP-EMTP can be made in a relatively simple manner, introducing actual transformer magnetizing characteristics. An example is shown in Figure 3.5, where the Type 93 "true nonlinear inductor" model in ATP-EMTP is applied, simulating actual 550 kV, 1 GVA transformer characteristics.[3] This inductor model is applied to introduce the initial residual flux. (For detailed data, see the file.[3]) In the data file applied in this case, magnetizing characteristics are simplified due to less detailed data being available. For correct calculations, more accurate characteristics might be necessary.

In Figure 3.5, the current gradually decays due to, mainly, the resistance in the circuit. The actual applied voltage to the inductance portion of the inductor impedance is reduced by the resistance drop, so the induced voltage is more or less asymmetrical due to the asymmetry of the magnetizing current. Therefore, the current gradually becomes a symmetrical one, down to less than 1 A. For correct calculation of the damping of the current, attention should be paid to the appropriate resistance value(s) in the circuit.

---

[3] ATPData3-03.dat, DAT3-03.acp: 550 kV 1000 MVA transformer inrush magnetizing current.

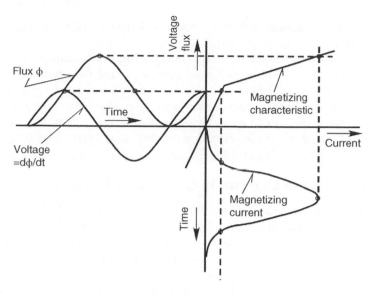

Figure 3.4   Transformer magnetizing.

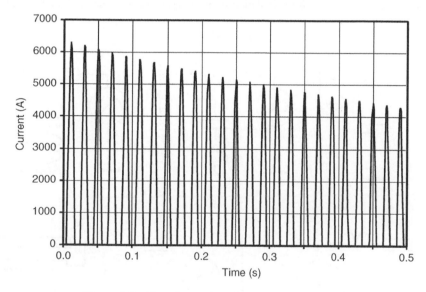

Figure 3.5   Transformer inrush magnetizing current.[3]

Notes:

- In the data file used for calculating Figure 3.5,[3] the magnetizing characteristic is represented by only three segments. If a more correct magnetizing current characteristic is desired, more detailed modeling for the current range of tens to several thousand amperes may be required.
- Three kinds of nonlinear inductor models are available:

- **Type 98 Pseudo nonlinear inductor**: Most simple and useful for general usage, but initial (residual) flux is not applicable.
- **Type 93 True nonlinear inductor**: Initial flux is applicable. Calculation speed is only a little bit slower. Care should be paid when applying initial flux; current does not start from zero but a certain value relevant to the flux value on the magnetizing curve.
- **Type 96 Pseudo-nonlinear hysteretic inductor**: Care should be taken when going up or down, the current/flux locus traces the same line for each, that is, the width between the two lines (up or down in current axis) is constant irrespective of the flux amplitude.

3.3 Transient Inrush Currents in Capacitive Circuits

Capacitive circuits, such as high-capacity shunt capacitor banks or EHV underground cable systems, when closed by relevant switching facilities (i.e., circuit breakers), may create a very high inrush current up to the same order of the short-circuit current, the transient frequencies of which are in the order of a few hundred hertz to several tens of kilohertz. The transient generally only lasts for a short time interval, so the consumption of the contact in the relevant switching facility (circuit breaker) is of most concern. Also, facilities are influenced by electromagnetic forces. Some examples of circuit diagrams are shown in Figure 3.6 in the single phase. Most actual circuits are in three phases, so for calculation in most cases, three-phase modeling is required.

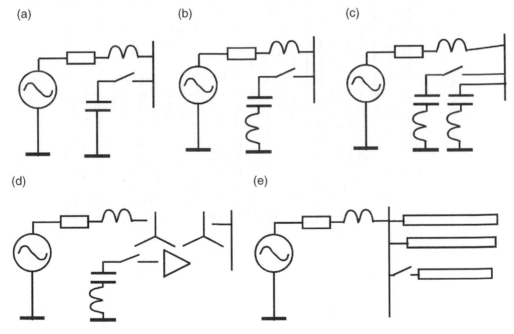

Figure 3.6 Some circuits creating inrush currents by capacitances. (a) Single capacitor bank. (b) Single capacitor bank with series reactor. (c) Back-to-back capacitor bank. (d) Capacitor bank in transformer tertiary winding circuit. (e) EHV cable system.

Figure 3.6 shows the following cases of circuits:

(a) Single capacitor bank circuit in the most simplified representation. The highest current is easily calculated by V (voltage), C (capacitance), and L (total series inductance). R (series resistance) affects only the damping of the transient current.
(b) Same as (a), but with the series reactor intentionally inserted. The practice is very common in Japan for suppressing the transient current and also the harmonic current component. In a so-called back-to-back capacitor bank circuit with a series reactor, a severe transient is safely and most efficiently damped.
(c) The so-called back-to-back shunt capacitor bank arrangement. When a capacitor bank is switched on while another one nearby was previously energized, true inrush current may flow. Each series impedance in the circuit, residual impedance in the capacitor itself, and so on must be carefully evaluated in calculations.
(d) In high-capacity substations capacitor banks are installed in the transformer's tertiary winding circuits. Thus, the amount of series reactances is automatically introduced.
(e) In EHV underground cable systems where plural circuits in particular are connected to the bus bar, a fairly high inrush current may be created.

For these cases, EMTP calculations are not thought to be quite so complicated, so no calculation example is shown here. Care should be taken as well for the damping elements (resistances) in the circuits. The values are to be based on the relevant (inrush current) frequencies, such as transmission lines, cables, transformers, and so on. It may be necessary to preliminarily calculate the frequency of the inrush current, and then recalculate after fixing the appropriate resistance value(s). The following are general ideas regarding damping, which might be of help unless otherwise obtained.

- **Overhead transmission line and underground cable**: Parameters are to be calculated based on the relevant transient frequency. For underground cable, dielectric loss (tanδ), which can be neglected in power frequency, might be necessary to count in. See Chapter 2.
- **Capacitor bank**: Appropriate dielectric loss based on the relevant frequency range is to be considered. The loss of the series reactor, if any, is in the order of 0.05% of the capacitor bank capacity in power frequency. About 60% of it is copper loss and can be represented by a constant value of series resistor irrespective the frequency. Iron loss (~25% in power frequency) is represented by a constant value of resistor connected in parallel due to the fact that the loss depends on the second power of the voltage irrespective of the frequency. Stray loss (~15% in power frequency) is proportional to the second power of current and 1.5th power of frequency.
- **Power transformer**: Typical high-capacity transformer losses are the following: Iron loss is about 0.03% of the capacity, which can be represented by a constant resistor connected in parallel to the magnetizing circuit. Load dependent loss is 0.15~0.2%, 85, and 15% of which are copper loss and stray loss, respectively. Like a capacitor bank's series reactor, the relevant losses are applied.

Appendix 3.A: Example of ATPDraw Sheets—Data3-02.acp

This file is used to calculate short-circuit current in a very high-capacity power system. The circuit diagram, corresponding to Figure 3.3, and an auto-transformer data inputting window are shown in Figure A3.1.

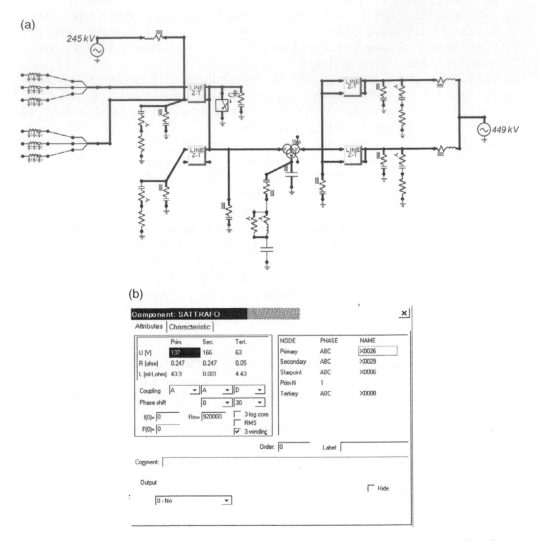

Figure A3.1 ATPDraw pictures in Data3-02.acp. (a) Power system circuit diagram. (b) Transformer data inputting window.

Reference

[1] E. Haginomori (1995) Short-circuit current distortions in power systems with high parallel capacitances, *IEEJ Transactions on Power and Energy*, **115-B**, 1, 85–90 (in Japanese).

4

Transient at Current Breaking

For circuit breakers or other switching facilities, successful current breakings depend on the competition between insulation recoveries and transient voltages, which are called transient recovery voltages (TRV) across the contacts just after the current breakings. The TRV appears as a circuit transient and can be calculated by simulating the circuit and breaking the current with the Electromagnetic Transients Program (EMTP). Nevertheless, the current injection principle gives simplified and effective calculation processes. The principle can be even used for obtaining a rough idea of TRV effectively.

In the current injection principle as shown in Figure 4.1, the current breaking condition (a) can be replaced by the condition of the inverse polarity of current $(-I)$, which is the condition (c), superimposed from the switch terminals to the originally flowing short circuit current (I), which is the condition (b). Condition (b) corresponds to the short-circuit condition where the TRV does not appear. The TRV that appears with condition (a) is same as the one with condition (c) due to the superimposing principle. In other words, the TRV appears between contacts after the current breaking is the same as the voltage that appears between contacts when the inverse polarity current of the prospective breaking current is injected into the circuit seen from the circuit breaker terminals. In condition (c), any initial condition is excluded. The method is only applicable for the voltage between breaker contacts. For other variables, for example, voltage to ground, the original process ((a) in Figure 4.1) is to be taken.

In the calculation, the ramp current instead of the sinusoidal one is mostly applicable due to a relatively short TRV duration time, although the breaking current is a sinusoidal one. The forced current breaking occurs before the natural current zero point at the breaking by a semiconductor device. In this case, the injection current of a step function is used.

Power System Transient Analysis: Theory and Practice using Simulation Programs (ATP-EMTP), First Edition.
Eiichi Haginomori, Tadashi Koshiduka, Junichi Arai, and Hisatochi Ikeda.
© 2016 John Wiley & Sons, Ltd. Published 2016 by John Wiley & Sons, Ltd.
Companion website: www.wiley.com/go/haginomori_Ikeda/power

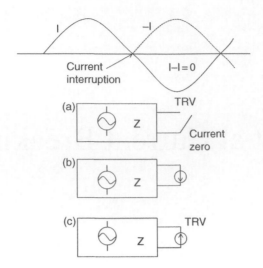

Figure 4.1   Current injection principle in current breaking.

4.1   Short-Circuit Current Breakings

The short-circuit current is the largest fault current. The TRV at breaking the short-circuit current changes according to the circuit impedances, which are composed of the reactor L, the capacitor C, and the resistor R. The distributed line parameter is also included. Typical examples are shown in Figure 4.2. Parts (1)–(5) in Figure 4.2 are a simple case. Parts (6)–(9) are a typical case in an actual system.

Programs made by using EMTP-ATP and ATPDraw are included in the data file[1] for all cases shown in Figure 4.2. Figure 4.3 shows the TRVs with the ramp currents calculated by ATPDraw for all cases in Figure 4.2.

In numerical calculations, finite values of increasing as a step are applied for the ramp current that might, initially, yield astonishing results for cases (2) and (4). To avoid the issue, it is recommended that small capacitances of 0.002 µF are connected in parallel to the inductances.

An analytical resolution can be obtained for a simple circuit. In the Laplace domain the voltage V and the current I are expressed by the following equations:

$$V = Z \times I(s)$$

$$I(s) = I_0 / s^2.$$

Z and V are shown in Table 4.1, in which Z_0 is the characteristic impedance of distributed parameter line.

The TRV after the short-circuit breaking can be understood by applying the current injection principle. The source-side and the line-side TRV can be obtained by using the current injection principle independently. The source-side TRV is represented by the circuit in Figure 4.2 (3) before the reflection at the line end comes by taking the line characteristic impedance as a resistor. The ITRV is inserted in this TRV calculated by the circuit of Figure 4.2 (9). After the reflection arrives, the circuit becomes (5), (6), or (7) of Figure 4.2. On the other hand, the line-side TRV is calculated by the circuit of Figure 4.2 (9).

---

[1] Data4-01: Current injection to various circuit elements.

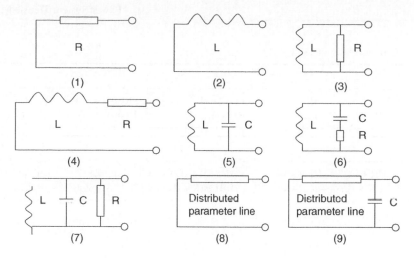

Figure 4.2 Some circuit elements to represent systems.

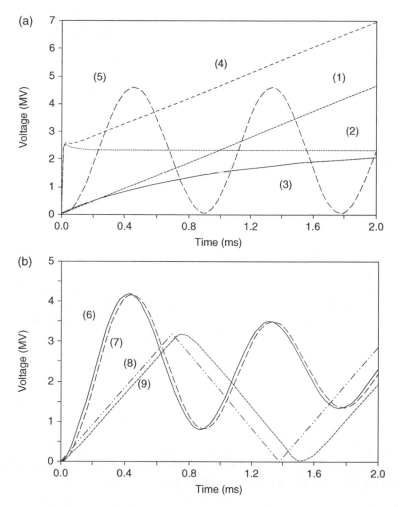

Figure 4.3 Calculated TRVs by injecting ramp currents to circuit elements in Figure 4.2. (a) Calculated TRV for Figure 4.2, from (1) to (5). (b) Calculated TRV for Figure 4.2, from (6) to (9).

Table 4.1 Equation of Z and V.

Case	1	2	3	4	5
Z	R	sL	sRL/(R+sL)	sRL/(R+sL)	sL/(s²LC+1)
V	I_0R/s^2	I_0L/s	I_0RL/s (R+sL)	I_0(sL+R)/s²	$I_0L/s(s^2LC+1)$

Case	6	7	8	9
Z	sL(sCR+1)/(1+sCR+s²LC)	sRL/(R+sL+s²LC)	Z_0	$Z_0/(1+sCZ_0)$
V	I_0L(sCR+1)/s(1+sCR+s²LC)	I_0RL/s(R+sL+s²LC)	$I_0 Z_0/s^2$	I_0Z_0/s^2 (1+sCZ₀)

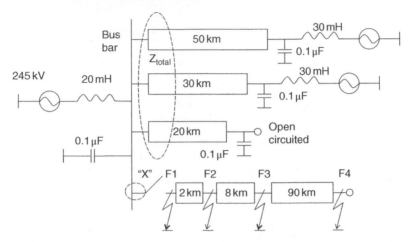

Figure 4.4 One-phase circuit diagram around a 300 kV substation to calculate TRVs in short-circuit breakings.

Next the current breaking is calculated for the actual system. Figure 4.4 shows a single-phase circuit diagram around the 300 kV substation. The circuit connected through the transformer is represented simply by a voltage source and an inductance equivalent to the transformer inductance and the system short-circuit reactance.

The transmission lines near the circuit breaker are relatively accurately represented by distributed parameter lines. On the other hand, remote systems are simply represented by lump elements such as capacitors and inductors. "X" is the connection bus bar relating to the initial transient recovery voltage (ITRV). The short-circuit current is 50 kA, which means a large capacity substation. Fault points are selected from F1 to F4. A fault at F1 is a so-called terminal fault, and faults at F2, F3, and F4 are all line faults. In particular, the fault at F2 is called a short line fault (SLF), which, due to the relatively high breaking current that is nearly 90% of the short-circuit current and the very high rate of rise of TRV due to the distributed line, is of importance for a certain type of circuit breaker.

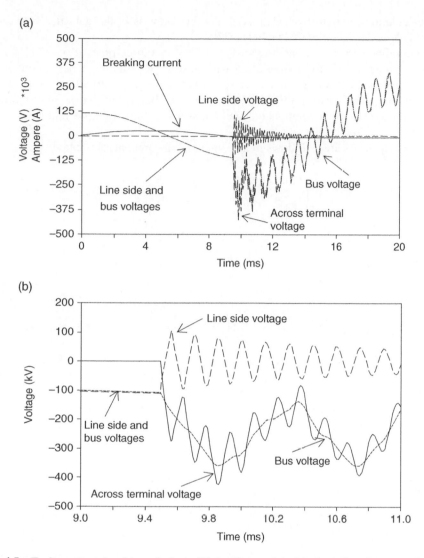

Figure 4.5 Fault current breaking, fault at F3 in Figure 4.4. (a) Overall voltages and current. (b) Enlargement around current interruption.

The calculation results are shown in Figure 4.5a,b for the case when the fault occurs at F3. Part (a) is the result during 40 ms and (b) is the result during 2 ms around the current breaking. Before the current breaking, a portion of voltage appears at the line side and bus side of the breaker, as shown by "line side and bus voltages" in Figure 4.5a. After the current breaking, the bus-side voltage converges to the source-side voltage with a transient. The line-side voltage becomes zero with a transient that has a particular triangle shape, as shown in Figure 4.5b.

The calculation program by EMTP-ATP and ATPDraw is in the data-file[2] that does not include ITRV and a data file[3] that includes ITRV.

As it would be very complicated to analyze a TRV in a three-phase circuit [1], it is practical to calculate a TRV by a simulation with an EMTP program. It is sometimes quite convenient for understanding qualitatively, in case of roughly estimating a rate of rise of recovery voltage. It is especially helpful to start out being familiar with the following results by the symmetrical components method, as it is extremely important when working with a circuit breaker.

The basic equations are as follows in the symmetrical components method:

$$
\begin{bmatrix} e_0 \\ e_1 \\ e_2 \end{bmatrix} = \begin{bmatrix} i_0 Z_0 \\ i_1 Z_1 \\ i_2 Z_2 \end{bmatrix} = \frac{1}{3} \begin{bmatrix} 1 & 1 & 1 \\ 1 & a & a^2 \\ 1 & a^2 & a \end{bmatrix} \begin{bmatrix} e_u \\ e_v \\ e_w \end{bmatrix} \tag{4.1}
$$

$$
\begin{bmatrix} i_0 \\ i_1 \\ i_2 \end{bmatrix} = \frac{1}{3} \begin{bmatrix} 1 & 1 & 1 \\ 1 & a & a^2 \\ 1 & a^2 & a \end{bmatrix} \tag{4.2}
$$

$$
a = \frac{1}{2} \left(-1 + j\sqrt{3} \right), \tag{4.3}
$$

where e_u, e_v, and e_w are voltages appearing across the terminals of the circuit breaker. Notations u, v, and w relate to phases. Z_0, Z_1, and Z_2 are respective sequence impedances seen through circuit breaker terminals; e_0, e_1, and e_2 are voltages; and i_0, i_1, and i_2 are currents.

In most cases of power transmission systems, $Z_1 = Z_2$. Z_2 is replaced by Z_1. The TRV is obtained analytically for the condition where the first phase current is broken after the three phases are earthed and short-circuited at one of the breaker terminals. e_v and e_w are zero before those phases are broken. The basic equation is

$$
\frac{1}{3} \begin{bmatrix} e_u \\ e_v \\ e_w \end{bmatrix} = \frac{1}{3} \begin{bmatrix} Z_0 i_u & Z_0 i_v & Z_0 i_w \\ Z_1 i_u & a Z_1 i_v & a^2 Z_1 i_w \\ Z_2 i_u & a^2 Z_2 i_v & a Z_2 i_w \end{bmatrix}. \tag{4.4}
$$

Regarding i_u and e_u, the following equations are derived:

$$
i_u = \frac{\begin{bmatrix} e_u & Z_0 & Z_0 \\ e_u & a Z_1 & a^2 Z_1 \\ e_u & a^2 Z_2 & a Z_2 \end{bmatrix}}{\begin{bmatrix} Z_0 & Z_0 & Z_0 \\ Z_1 & a Z_1 & a^2 Z_1 \\ Z_2 & a^2 Z_2 & a Z_2 \end{bmatrix}}. \tag{4.5}
$$

---

[2] Data4-02: TRV example calculation.
[3] Data4-03: TRV calculation, including ITRV—current injection is applied for TRV calculation.

The equivalent impedance for the first pole to clear is the ratio of e_u and i_u:

$$\frac{e_u}{i_u} = \frac{3Z_0 Z_1}{\left(2Z_0 + Z_1\right)}. \tag{4.6}$$

Likewise, for the second and third pole to clear, the following are introduced, respectively:

$$\frac{e_v}{i_v} = \frac{\left(2Z_0 Z_1 + Z_1^2\right)}{\left(Z_0 + 2Z_1\right)} \tag{4.7}$$

$$\frac{e_w}{i_w} = \frac{\left(Z_0 + 2Z_1\right)}{3}. \tag{4.8}$$

TRVs are conceptually considered as products of injection currents and equivalent impedances in a three-phase circuit as well. Therefore, TRVs in three-phase circuits can be guessed, at least for relative values or qualitatively from the previously derived impedance values. For quantitatively accurate values, of course, EMTP calculations are inevitable.

4.2   Capacitive Current Switching

The current of charging capacitances such as no-load overhead transmission lines, underground cables, or shunt capacitor banks is called capacitive current. Figure 4.6 shows a simplified circuit for a capacitive current breaking and the change of a voltage and current during breaking. At the breaking contacts of a circuit breaker, a switching device is opened, after which the current is broken and the current reaches a zero point.

The phase of a capacitive current and that of the voltage of capacitance differ by about 90°. At the current zero point, the voltage of the capacitance is at a maximum value that remains on the capacitance after the current breaking. After a time duration corresponding to 180 electrical degrees, the source voltage polarity is reversed and the voltage between contacts becomes higher than twice that of the source voltage, as shown in Figure 4.6. The breakdown may happen between contacts due to such high voltage, as it is known by the term "restrike." Afterward, the breakdown may be extinguished and the breakdown

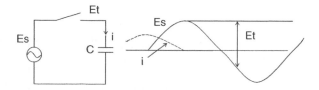

Figure 4.6   Capacitive current breaking—most simplified representation.

and extinction may be repeated, which results in a gradual voltage escalation. For modern sophisticated power systems with reduced insulation levels, restrike-free is a serious requirement.

4.2.1   Switching of Capacitive Current of a No-Load Overhead Transmission Line

The switching of capacitive current of a no-load overhead transmission line occurs when a circuit breaker located at a line end is operated with another line end opened [2]. The switching occurs relatively frequently in a power system. The operation called "rapid auto reclosing" is adopted for a transmission line. When an overhead transmission line suffers from a lightning overvoltage, an arc can bridge several lines, which results in a fault current flowing by a short circuit and/or ground faults. Circuit breakers at both ends of the faulted transmission line are broken and a short-circuit current arc is extinguished on the fault overhead transmission line. After 0.3–1 s, when the insulation of arc-generated area is recovered, the circuit breakers are closed to restart a power transmission through the line. The period of an electricity outage or a voltage drop at a parallel line operated is reduced and the quality and service of electricity supply is expected to improve.

For a rapid auto reclosing there are two schemes. One of them is a three-phase reclosing in which three phases all are reclosed simultaneously; another is a single-phase one in which only a fault phase is reclosed. In the three-phase reclosing a closing makes an overvoltage as high as a restrike, since a switching of the line without a fault is a capacitive current one, and the circuit breaker is closed with a residual voltage or charge on a line, depending on a closing timing. Overhead transmission lines operated in electric power systems below 275 kV of nominal voltage are designed to endure the overvoltage. The overhead transmission line of 550 kV of maximum voltage is designed on the condition that the reclosing overvoltage is suppressed by a countermeasure, which is practically a two-step closing, by coupling a circuit breaker with a parallel resistor. The analysis of phenomena related to a transmission line is very complicated due to conditions where there are several conductors connected electromagnetically and electrostatically, there are influences from an earth and a skin effect of conductors in a high-frequency domain, and there is occasionally an attenuation aspect from a corona discharge to be considered. The numerical calculation by using the EMTP program is a smart solution, as EMTP was mainly developed to solve overvoltage phenomena of transmission lines.

The system layout for a 550 kV of maximum voltage and 150 km lines is shown in Figure 4.7 for a calculation of various switching phenomena. The calculation results are shown in Figure 4.8a,b. The first case in (a) shows that the maximum voltage between contacts reaches 2.32 times of the source voltage peak value in a three-phase switching due to the electrostatic coupling of conductors. The maximum voltage in a single-phase breaking is less than 2. The second case (b) is the result where the breaking timing of the phase in which a current zero point comes after the first phase breaking is delayed for some reason. The voltage between contacts reaches up to 2.8 times higher than a source voltage peak value. Details are shown in the data files.[4,5]

---

[4] Data4-11: 550 kV line normal breaking.
[5] Data4-12: 550 kV line capacitive current breaking—second phase delayed.

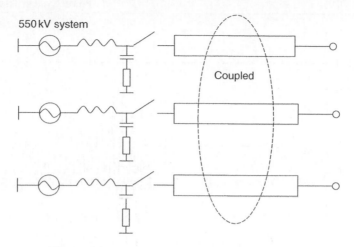

Figure 4.7 System layout of no-load overhead line.

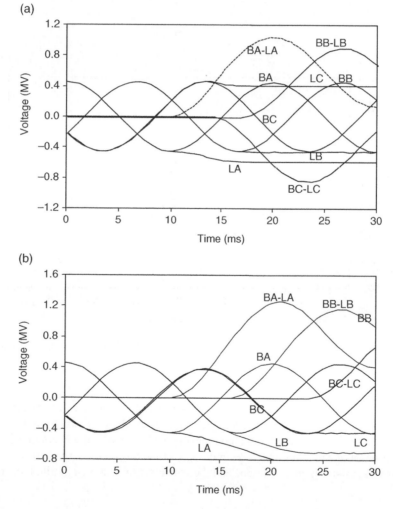

Figure 4.8 Calculation result of no-load overhead line. (a) Normal breaking. (b) Delaying in second pole to clear.

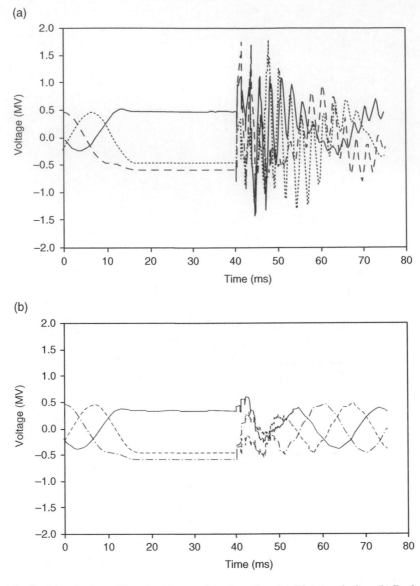

Figure 4.9   Rapid reclosing with and without resistor insertion. (a) Direct reclosing. (b) Reclosing with resistor insertion.

The last case, illustrated in Figure 4.9, is related to the rapid auto reclosing. At a most severe overvoltage, the calculation is conducted for a three-phase opening without a fault and reclosing. In case (a), where the rapid auto reclosing is done without any countermeasure, the maximum overvoltage reaches up to 4.2 times higher than the source voltage peak. In case (b), where a 1000 Ω resistor is inserted once and then the resistor is short-circuited, the maximum overvoltage is suppressed down to 1.6 times. For details of the system and operation sequence parameters, see the data files.[6,7]

4.2.2   Switching of Capacitive Current of a Cable

Figure 4.10 shows capacitive current breaking in a cable system. The cable is modeled as a screened one, that is, each phase core is surrounded by an earthed screen so that no electrical static coupling exists between phase cores, corresponding to equal zero and positive sequence capacitance values. The supply side is modeled as a non-earthed neutral condition.

In EMTP, one-terminal-grounded voltage source is mandatory. For representing a non-earthed source, a combination of a current source and impedance can be applied. Alternatively, the semi-ideal transformer or "No. 18 ungrounded source" can also be applied.

Calculation results are shown in Figure 4.11. The result (a) is a usually specified case for a nonsolidly earthed neutral system, and due to the enhancement (shifting up) of the supply-side neutral voltage, the maximum recovery voltage reaches up to 2.5 times that of the source voltage. In (b), as more general cases, significant values of capacitances to ground, such as cables, are connected to the supply-side bus bar. Then due to less enhancement of the neutral voltage in the supply side, the maximum recovery voltage is approximately 2.0 times as much reduction is expected (also see the data files[8,9]).

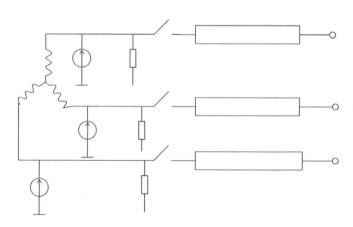

Figure 4.10   Circuit diagram for capacitive current breaking in a system with a non-solidly earthed source circuit.

---

[6] Data4-13: 550 kV line, reclosing at relevant timings creating max overvoltages.
[7] Data4-14: 550 kV line, reclosing with resistor insertion.
[8] Data4-15: Cable capacitive current breaking—no significant to-ground capacitance in the source circuit.
[9] Data4-16: Cable capacitive current breaking—significant to-ground capacitance (cables) connected in source.

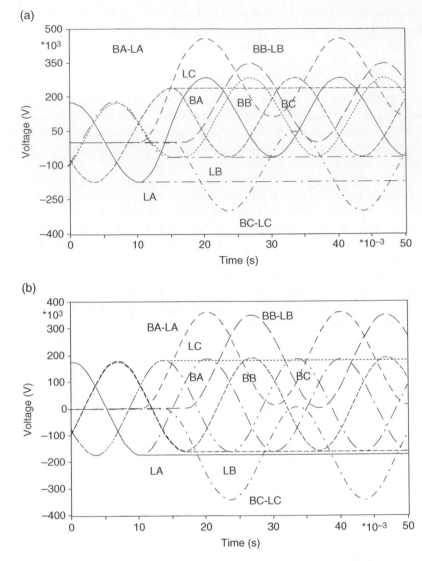

Figure 4.11 Calculation results for capacitive current breaking in system with a non-solidly earthed source circuit. (a) Isolated neutral source circuit to earth. (b) With significant capacitance in source circuit.

4.2.3   Switching of Capacitive Current of a Shunt Capacitor Bank

A capacitor bank has been introduced significantly in a power system mainly for improving system stability. Switching of the bank is conducted frequently according to variations of load. An example of breaking shunt capacitor bank capacitive current, with a 66 kV and 50 MVA rating, is shown in Figure 4.12, in which (a) is a circuit diagram and (b) is a calculation result. The supply circuit neutral is grounded through a resistor with a high ohm value. In the calculation, No. 18 ungrounded source in EMTP menu is applied. The capacitors have in some

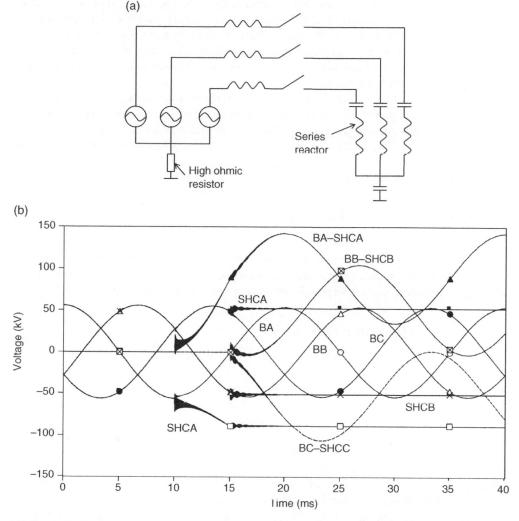

Figure 4.12 Shunt capacitor bank capacitive current breaking-66 kV, 50 MVA bank. (a) Circuit diagram. (b) Voltage changes in breaking.

cases series-connected reactors, the purpose of which is to suppress harmonics (higher than third stage) and back-to-back inrush-making currents. (Also see the data file.[10])

There are many cases in which the series reactor is not equipped. When an additional capacitor bank is connected to an already operating capacitor bank, a very large inrush current flows between two capacitor banks. The phenomenon is called "back to back."

The phenomenon at breaking is complicated due to a three-phase condition, and a numerical calculation by an EMTP program is recommended.

---

[10] Data4-17: Shunt capacitor bank circuit, 66 kV to 50 MVA—No. 18 source representing isolated neutral source circuit applied.

The voltage charged on the capacitor is enhanced by 6% due to the inverse polarity of voltage on the series-connected reactor. As the enhanced voltage remains on the capacitor as DC voltage after the breaking, the recovery voltage between contacts is also enhanced by the value. The recovery voltage is 2.7 p.u. in this case, while it is 2.5 p.u. without the reactor. The rated voltage of capacitor bank increases by 6% and an additional reactor is necessary. The total cost of capacitor bank is high.

Moreover, due to the voltage oscillation on the reactor, a high-frequency component is involved at the initial part of the recovery voltage. When a restrike occurs during the capacitor bank breaking, an extremely large charged energy oscillates in a circuit, which influences dielectric performance and electromagnetic force a lot. As a relatively high frequency of oscillation is created by the series-connected reactor, in an EMTP calculation, a sufficiently low step time value is to be used.

4.3   Inductive Current Switching

The inductive current is one with almost 90° behind the voltage, such as a shunt reactor current, a transformer magnetizing current, and a stalled motor energizing current. When the inductive current is broken, various specific phenomena occur. As a result, an overvoltage, which gives excessive electric stresses to equipment, will be generated.

4.3.1   Current Chopping Phenomenon

When the inductive current is broken, the "chopping" phenomenon often occurs. This term comes from the sudden stopping of current. Today's high-resolution measuring and analyzing methods have clearly explained the phenomenon: it occurs due to the current being broken at the current zero point generated by the superposition of high-frequency current [3]. This phenomenon only occurs with an SF6 gas circuit breaker and an air blast circuit breaker. With a vacuum circuit breaker, the phenomenon in which a current is suddenly forced to be zero can occur.

The chopping phenomenon is recognized as one in which the AC current is broken at the current zero point generated on the AC current by the high-frequency oscillating current enlarged due to the negative characteristics of resistance of arc in the circuit breaker. The current chopping, which is a forced current breaking before current zero point, is simulated by the usual switch element with a time control function. Figure 4.13a shows the circuit diagram of breaking the shunt reactor current at the rating of 300 kV and 150 MVA. The single phase is used for simplification. The inductance of the connection bus bar is located near the circuit breaker. Figure 4.13b shows voltages during the chopping and following phenomena.

The oscillation before the current chopping occurs only on the connection lines adjacent to the circuit breaker. The reactor current does not oscillate. Therefore, a whole magnetic energy corresponding to the chopping current value is stored in the reactor coil. The electrostatic energy corresponding to the reactor voltage is stored in the capacitance connected in parallel to the reactor. When the sum of two energies converses to the electrostatic energy, the voltage between capacitor terminals, that is, the voltage applied to the reactor coil, is at the maximum as an overvoltage. This can be written by the equation

$$\frac{1}{2}CV^2 = \frac{1}{2}Li^2 + \frac{1}{2}CV_0^2, \tag{4.9}$$

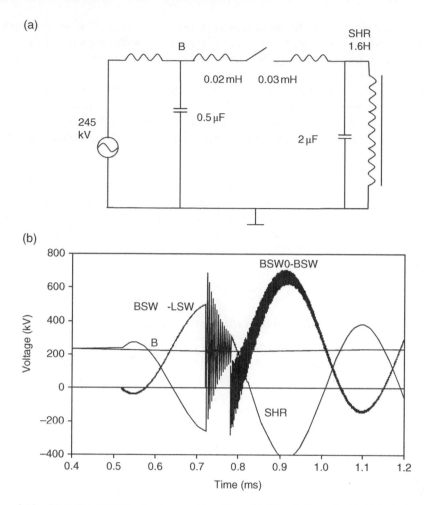

Figure 4.13 300 kV, 150 MVA shunt reactor breaking. (a) Circuit diagram. (b) Calculation result.

where V = maximum overvoltage, i_c = chopping current, and V_0 = voltage just before chopping, which is nearly equal to supply voltage.

The typical chopping current value of an existing circuit breaker is known as 5–50 A. And it is known as well to be proportional to the parallel capacitance. From the equation it is understood that the overvoltage will not be influenced by the parallel capacitance value. The larger the reactor inductance becomes, or in other words, the smaller the reactor capacity becomes, the larger the overvoltage becomes.

4.3.2 Reignition

The voltage between circuit breaker contacts increases relatively slowly after the current breaking, as shown by B-SHR in Figure 4.13b. When the gap length between circuit breaker contacts is not sufficient, the breakdown often occurs. The phenomenon is called reignition.

When a reignition occurs, charges on capacitances next to the small inductances at both sides of and adjacent to the circuit breaker go back and forth frequently through the small inductances. The overvoltage of very high frequency up to several hundred kilohertz is applied to the reactor coil terminals.

The overvoltage of such a high frequency is concentrated to the coil turns near the reactor terminal when it is applied to coils of a reactor, a transformer, and a motor. The overvoltage due to reignition is lower than the overvoltage due to chopping. The high frequency enhances the electric stress that cannot be rendered by a surge arrester, which can reduce the amplitude of oscillation.

The most influential stress is the high-frequency overvoltage due to reignition for coils of a transformer and a motor. Countermeasures, such as strengthening the coil insulation and minimizing reignition occurrence by improving the circuit breaker performance and adopting the contact separation timing control, have been used for preventing the coil insulation failure and keeping a reliability of the system especially for systems with the extremely high rated voltage 300 kV and the super-high rated voltage 550 kV. (Also see the data file.[11])

4.3.3   High-Frequency Extinction and Multiple Reignition

The current flowing in the circuit breaker due to reignition has a high amplitude of several kiloamperes at the maximum and frequency reached is up to several hundred kilohertz. The steepness of current decrease at the current zero point is so high that the circuit breaker cannot break it generally. However, it can attenuate rapidly and can be broken when its amplitude is attenuated to several hundredths. The electromagnetic energy is stored in a coil and the overvoltage generated is very likely due to chopping. The stored energy can be larger than that from chopping when special conditions are fulfilled. This means the higher overvoltage will be generated.

Due to the higher overvoltage from reignition than from chopping, the reignition might be repeated. This is called multiple reignitions. The occurrence of multiple reignitions is very rare. When it happens once, a very severe overvoltage is applied to the coils. Multiple reignitions depend on the breaking of the reignition current that is greatly influenced by the circuit condition of system. The condition is given and cannot be changed.

Multiple reignition produces a more severe overvoltage that cannot be controlled when it occurs once. It is thought that the best way to prevent the occurrence of a multiple reignition might be to prevent the occurrence of reignition, which is a base for all processes.

4.4   TRV with Parallel Capacitance in SLF Breaking

The TRVs are analyzed for various electric power source systems by using the current injection method in Section 4.1. The fault points F2 and F3 represent the SLF breaking conditions, where the triangle shape TRV appears, as shown in Figure 4.5. The TRV in the conditions can

---

[11] Data4-18:300 kV, 150 MVA shunt reactor current breaking—current chopping—reignition—HF current interruption.

be analyzed with the circuit of (8) in Figure 4.2 by using the current injection method. The condition with parallel capacitance corresponds to (9) of Figure 4.2.

A mathematical solution can be easily and simply obtained in a TRV calculation with parallel capacitance when a TRV without parallel capacitance is obtained. The TRV produced from a transmission line of an ideal distributed parameter has a triangle shape, one cycle of which can be expressed by

$$\mathrm{TRV}(s) = \frac{\omega I Z}{s^2}\left(1 - 2e^{-t_L s}\right), \tag{4.10}$$

where t_L is time to peak without parallel capacitance, ω is angular frequency of breaking current, I is breaking current, Z is characteristic impedance or surge impedance, and s is the Laplace denominator. Equation (4.10) is correct from time zero to $2t_L$. During this period, Equation (4.10) can be transformed by the following:

$$\mathrm{TRV}(s) = \frac{\omega I Z}{s^2}\left(1 - 2e^{-t_L s} + 2e^{-2t_L s} - \cdots + \cdots\right). \tag{4.11}$$

When there is attenuation on the triangle wave shape, the term $2e^{-t_L s}$ is modified to $2ke^{-t_L s}$ in Equations (4.10) and (4.11). Here, $k < 1.0$ is expected.

A TRV is a product of a breaking current and impedance in the Laplace domain as well. In a very short time period the injection current can be expressed in the Laplace domain as follows,

$$\frac{\omega I}{s^2}, \tag{4.12}$$

which corresponds to $\omega I t$ in the time domain.

Consequently, the transmission line with a distributed parameter can be expressed in Equation (4.13) in the Laplace domain:

$$Z\left(1 - 2e^{-t_L s}\right). \tag{4.13}$$

In the Laplace domain the lumped capacitance is expressed by Equation (4.14),

$$\frac{1}{Cs}, \tag{4.14}$$

where C is capacitance. The value of C contains one that generates the inherent delay time t_{dL} and one that will be added in a system.

As the line and the capacitance are connected in parallel, the total impedance becomes the following Equation (4.15):

$$\frac{Z\left(1 - 2e^{-t_L s}\right)}{Cs\left(\dfrac{1}{Cs} + Z - 2Ze^{-t_L s}\right)} = \frac{Z\left(1 - 2e^{-t_L s}\right)}{\left(1 + t_{dL}s\right) - 2t_{dL}se^{-t_L s}}, \tag{4.15}$$

where $t_{dL} = ZC$.

The TRV with parallel capacitance is a product of current in Equation (4.12) and the impedance in Equation (4.15). It can be expressed in Equation (4.16):

$$
\begin{aligned}
\omega IZ \times \frac{1}{s^2} &\times \frac{Z\left(1-2e^{-t_L s}\right)}{\left(1+t_{dL}s\right)-2t_{dL}se^{-t_L s}} \\
&= \omega IZ\left[\frac{1}{s^2}\times\frac{1}{1+t_{dL}s}-\frac{1}{s^2}\times\frac{2e^{-t_L s}}{\left(1+t_{dL}s\right)^2}+\cdots\cdots\right] \\
&= \omega IZ\left[\frac{1}{s^2}-\frac{t_{dL}}{s}+\frac{t_{dL}^2}{1+t_{dL}s}-2e^{-t_L s}\left(\frac{1}{s^2}-\frac{2t_{dL}}{s}+\frac{2t_{dL}^2}{1+t_{dL}s}+\frac{t_{dL}^2}{\left(1+t_{dL}s\right)^2}\right)\right].
\end{aligned}
\tag{4.16}
$$

The second part of Equation (4.16) is valid for $t_L < t < 2t_L$ only.

The TRV with parallel capacitance for the SLF breaking can be expressed in the time domain as follows. In the case of $0 < t < t_L$,

$$
TRV_1(t) = \omega IZ\left[t+t_{dL}\left(e^{-\frac{t}{t_{dL}}}-1\right)\right].
\tag{4.17}
$$

In the case of $t_L < t < 2t_L$,

$$
TRV_2(t) = TRV_1(t)-2\omega IZ\left[t'\left(e^{-\frac{t'}{t_{dL}}}+1\right)+2t_{dL}\left(e^{-\frac{t'}{t_{dL}}}-1\right)\right],
\tag{4.18}
$$

where $t' = t - t_L$.

In the case of $t > 2t_L$, Equation (4.11) is used instead of Equation (4.10). When attenuation is considered, the term $2ke^{-t_L s}$ should be used instead of the term $2e^{-t_L s}$ in Equation (4.10). The calculation processes will be changed slightly.

With or without capacitance, the TRV peak value is equal to ωIZt_L commonly. The TRV peak will not be attenuated significantly. By dividing Equations (4.17) and (4.18) by the peak value ωIZt_L, the following equations, (4.17a) and (4.18a), can be derived

$$
TRV(10) = \frac{t_{dL}}{t_L}\left[\frac{t}{t_{dL}}+\left(e^{-\frac{t}{t_{dL}}}-1\right)\right]
\tag{4.17a}
$$

$$
TRV(20) = TRV(10)-\frac{2t_{dL}}{t_L}\left[\frac{t'}{t_{dL}}\left(e^{-\frac{t'}{t_{dL}}}+1\right)+2\left(e^{-\frac{t'}{t_{dL}}}-1\right)\right].
\tag{4.18a}
$$

The results obtained by Equations (4.17a) and (4.18a) are all normalized by the peak value ωIZt_L and the delay time t_{dL} (Figure 4.14).

(a)

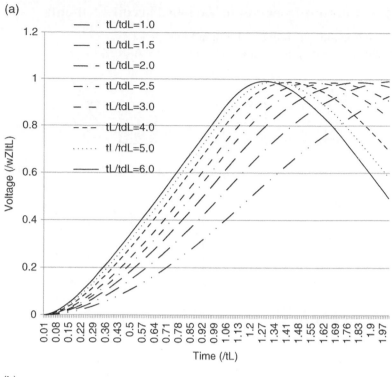

(b)

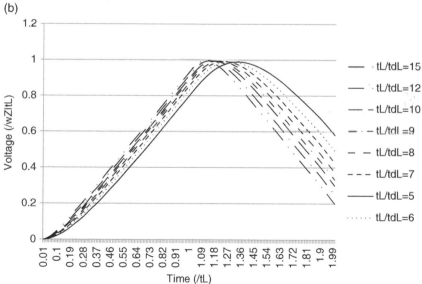

Figure 4.14 TRV with parallel capacitance normalized by the peak value ωIZt_L and the delay time t_{dL}. (a) $1 < t_L/t_{dL} < 6$. (b) $5 < t_L/t_{dL} < 15$.

Appendix 4.A: Current Injection to Various Circuit Elements

The circuit diagrams in the data file[1] are shown in Figure A4.1.

The data for the circuit components are shown in Figures A4.2–A4.12.

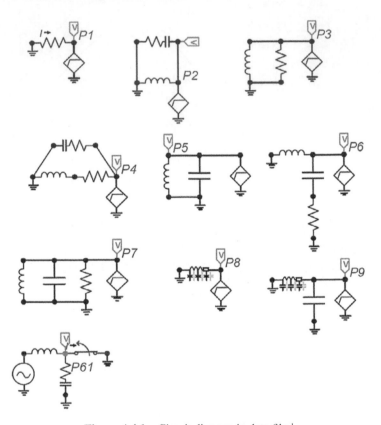

Figure A4.1 Circuit diagram in data file.[1]

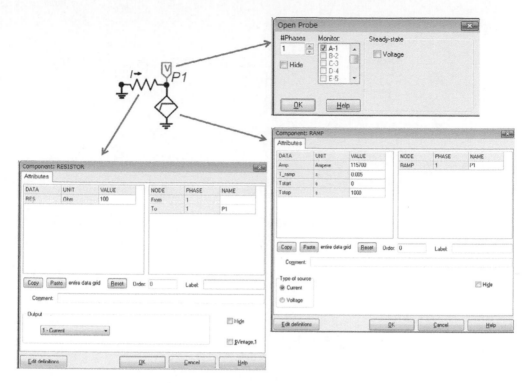

Figure A4.2 Circuit components data (1).

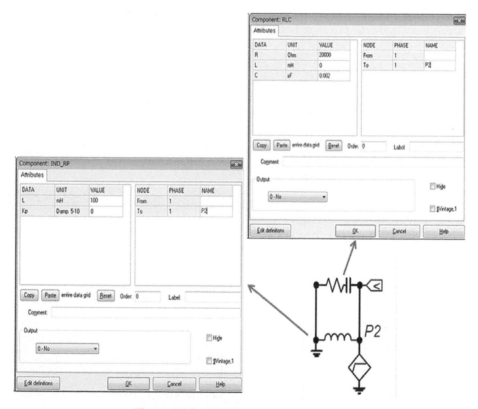

Figure A4.3 Circuit components data (2).

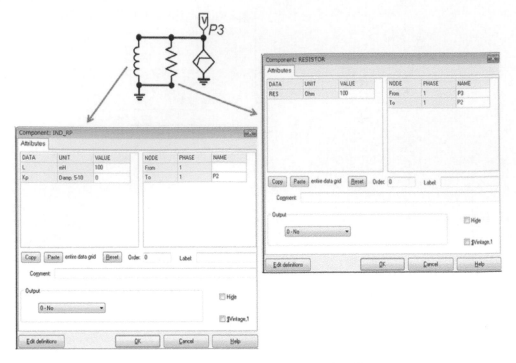

Figure A4.4 Circuit components data (3).

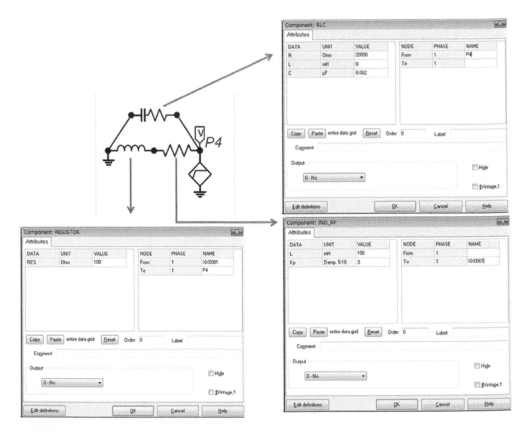

Figure A4.5 Circuit components data (4).

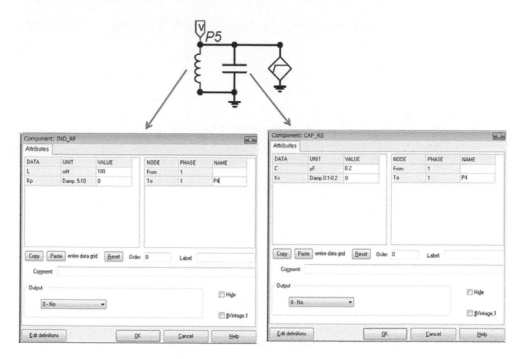

Figure A4.6   Circuit components data (5).

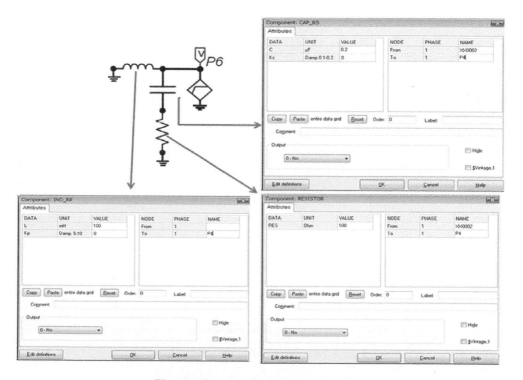

Figure A4.7   Circuit components data (6).

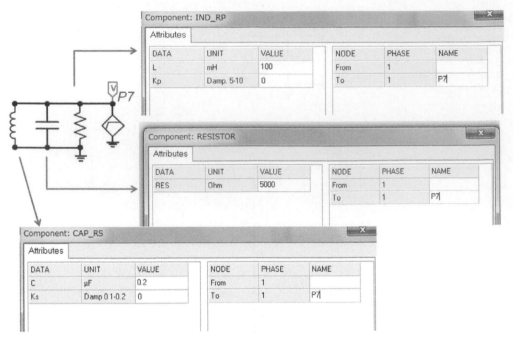

Figure A4.8   Circuit components data (7).

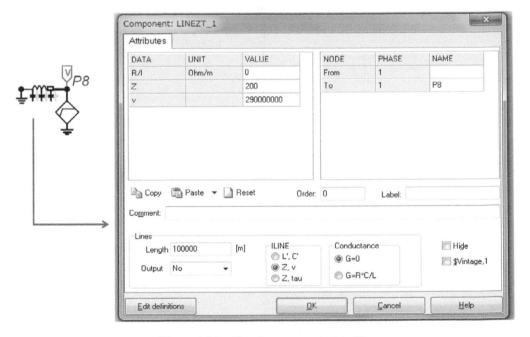

Figure A4.9   Circuit components data (8).

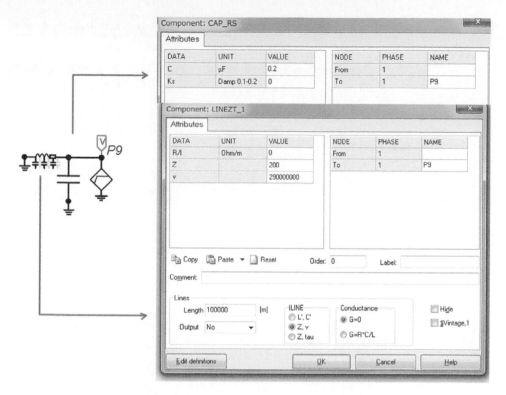

Figure A4.10 Circuit components data (9).

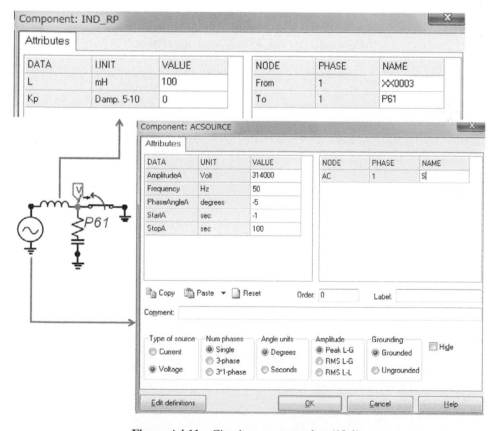

Figure A4.11 Circuit components data (10.1).

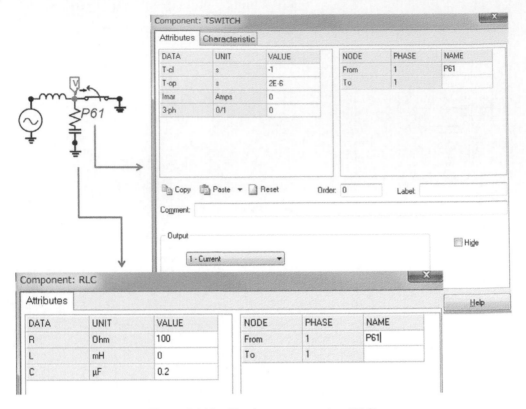

Figure A4.12   Circuit components data (10.2).

Appendix 4.B: TRV Calculation, Including ITRV—Current Injection is Applied for TRV Calculation

The circuit diagrams in the data file[3] are shown in Figure B4.1.

The data for the circuit components are shown in Figures B4.2–B4.11.

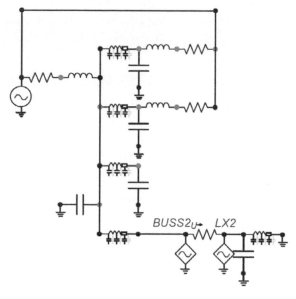

Figure B4.1   Circuit diagram in data file.[3]

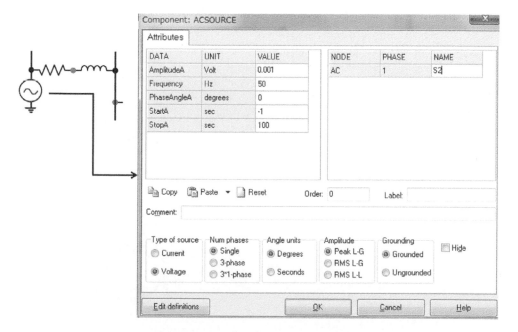

Figure B4.2   Circuit components data (1).

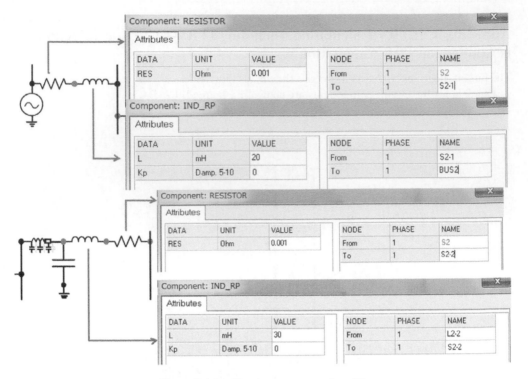

Figure B4.3   Circuit components data (2).

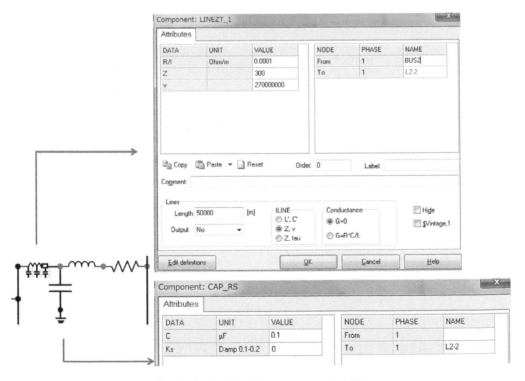

Figure B4.4   Circuit components data (3).

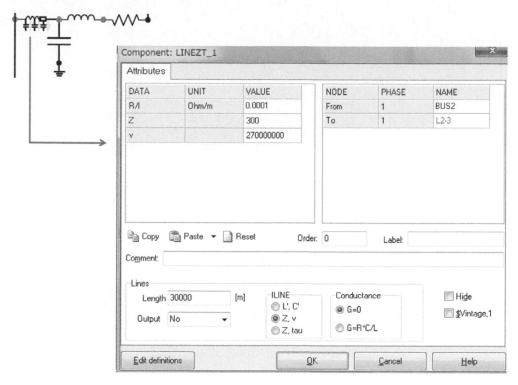

Figure B4.5 Circuit components data (4).

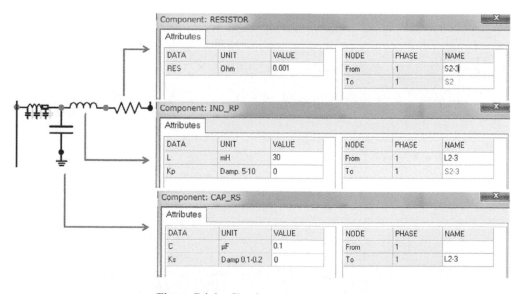

Figure B4.6 Circuit components data (5).

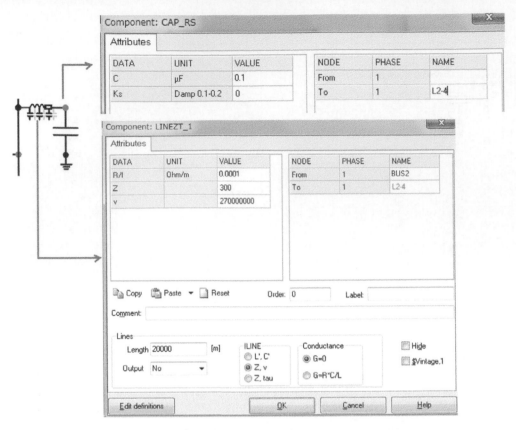

Figure B4.7   Circuit components data (6).

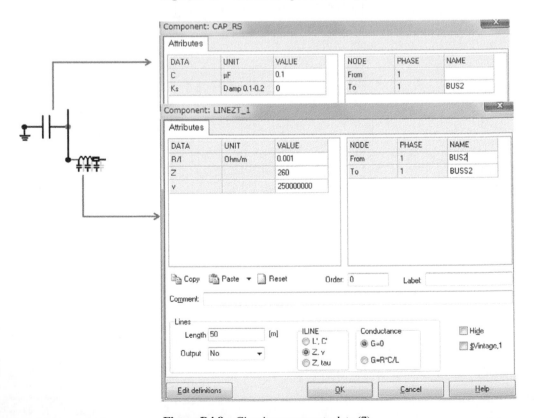

Figure B4.8   Circuit components data (7).

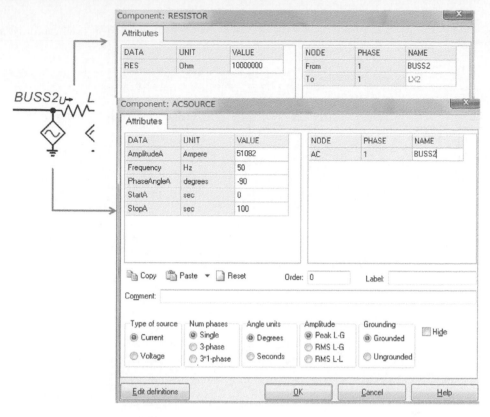

Figure B4.9 Circuit components data (8).

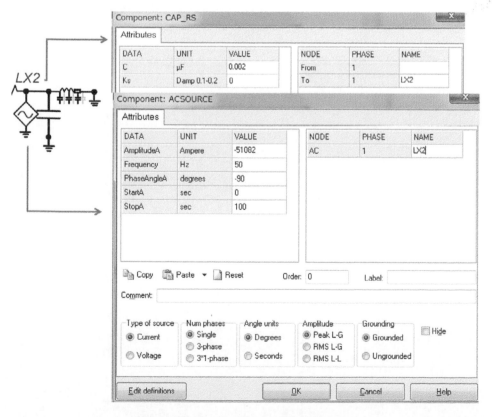

Figure B4.10 Circuit components data (9).

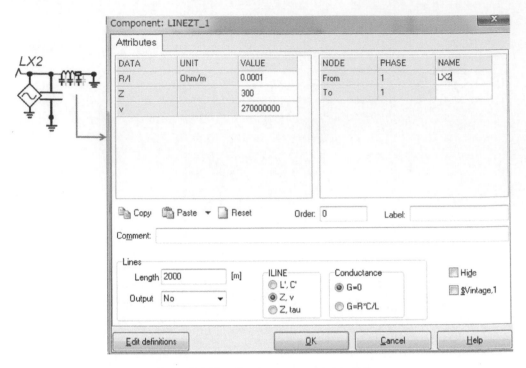

Figure B4.11   Circuit components data (10).

Appendix 4.C: 550 kV Line Normal Breaking

The circuit diagrams in the data file[4] are shown in Figure C4.1.
  The data for the circuit components are shown in Figures C4.2–C4.6.

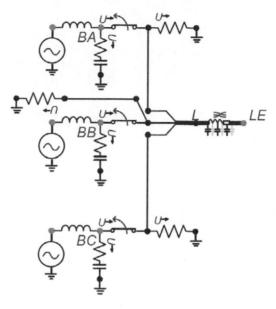

Figure C4.1 Circuit diagram in data file.[4]

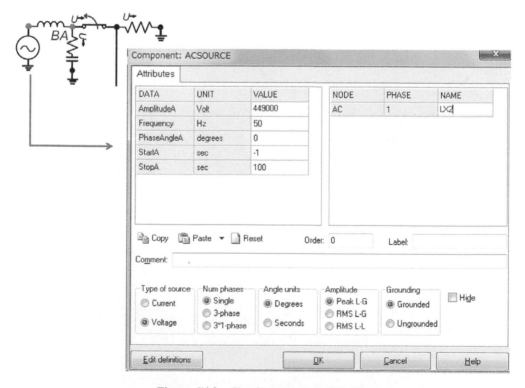

Figure C4.2 Circuit components data (1).

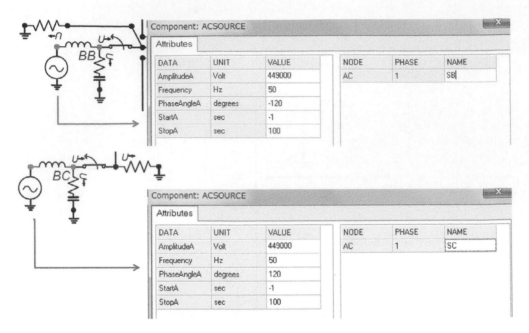

Figure C4.3   Circuit components data (2).

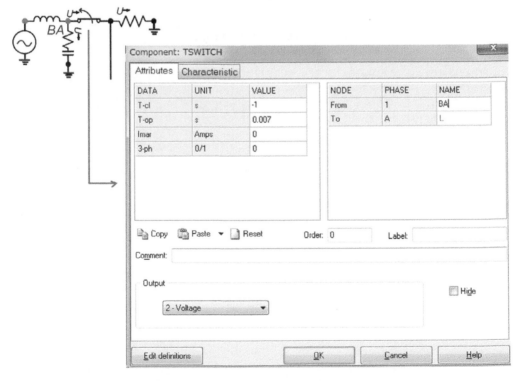

Figure C4.4   Circuit components data (3).

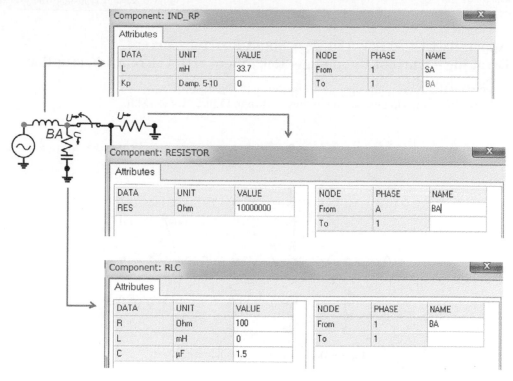

Figure C4.5 Circuit components data (4).

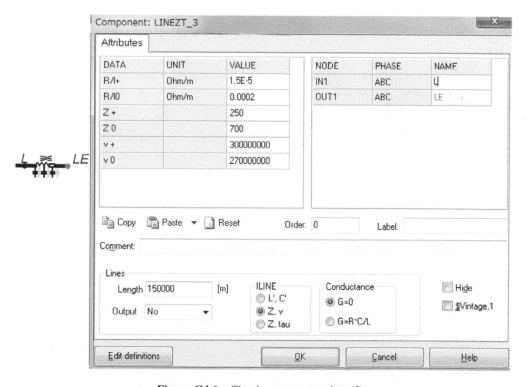

Figure C4.6 Circuit components data (5).

Appendix 4.D: 300 kV, 150 MVA Shunt Reactor Current Breaking—Current Chopping—Reignition—HF Current Interruption

The circuit diagrams in the data file[11] are shown in Figure D4.1.

The data for the circuit components are shown in Figures D4.2–D4.6.

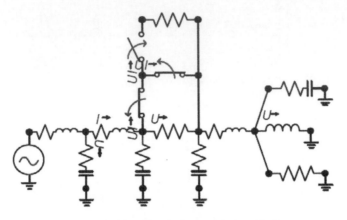

Figure D4.1   Circuit diagram in data file.[11]

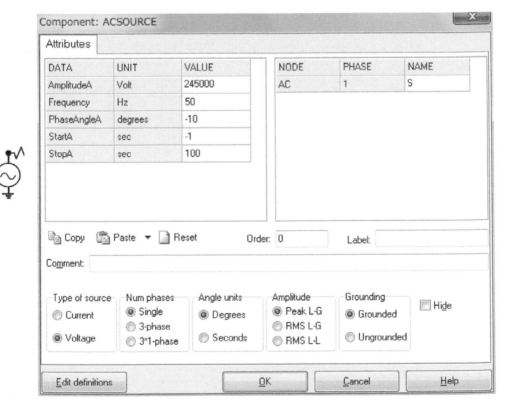

Figure D4.2   Circuit components data (1).

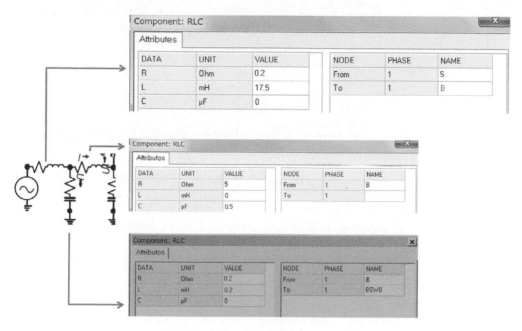

Figure D4.3   Circuit components data (2).

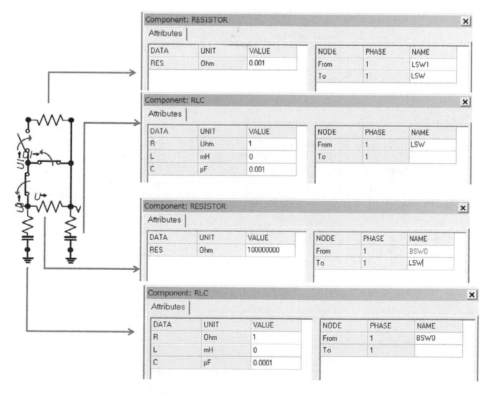

Figure D4.4   Circuit components data (3).

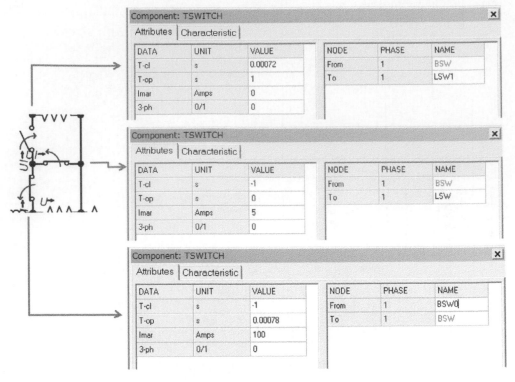

Figure D4.5 Circuit components data (4).

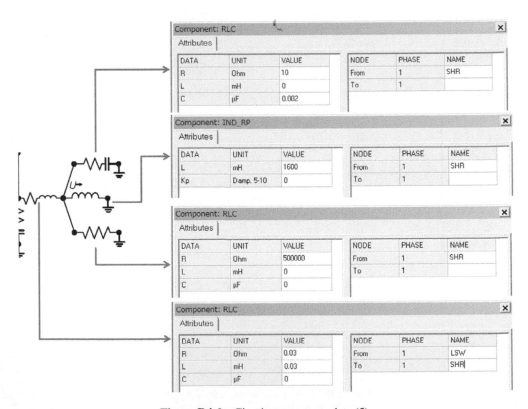

Figure D4.6 Circuit components data (5).

References

[1] M. Yamamoto, S. Yamashita, H. Ikeda, and S. Yanabu (1985) New short-circuit testing facilities to cope with the recent development of GIS, *IEEE Transactions on Power Apparatus and Systems*, **PAS-104** (1), 150–156.

[2] K. Suzuki, H. Toda, A. Aoyagi, H. Ikeda, A. Kobayashi, I. Ohshima and S. Yanabu (1993) Development of 550 kV 1-break GCB (PART I): investigation of interrupting chamber performance. *IEEE Transactions on Power Delivery*, **PWRD-8**, (7), 1184–1191.

[3] Y. Yamagata, S. Okabe, M. Ishikawa, H. Ikeda, S. Nishiwaki, Y. Murayama, and S. Yanabu (1988) High frequency current interruption at reactor switching by SF6 gas circuit breaker. *Proceedings of 9th International Conference on Gas Discharges and their Applications*, pp. 155–158.

5

Black Box Arc Modeling

In the previous chapter, breaking phenomena analysis was introduced using the arcing characteristic of a circuit breaker. It is expanded in this chapter to introduce breaking phenomena analysis by considering circuit breaker performance. It becomes sufficiently possible by using the recent Electromagnetic Transients Program (EMTP) to analyze an electric power system circuit combined with an arc expressed by electric resistance, which is derived as a solution from a differential equation regarding an arcing characteristic with variables such as a current, a voltage, and a time.

In order to understand phenomena in general related to the breaking performance of a circuit breaker, equations of the Mayr arc model and the Cassie arc model have been widely utilized for years. Special features and strong points of the two models are the following:

Mayr arc model: The Mayr arc model represents well the arcing characteristics of SF6 gas and air circuit breakers at a relatively small current below several tens of amperes, especially during the "interaction interval" of a few microseconds around a current zero point of a breaking current. Consequently, the model is frequently used to investigate breaking success and failure. Several attempts have been made to modify and expand the equation in order to express an arc characteristic with a high accuracy.

Cassie arc model: The Cassie arc model represents well the arcing characteristics of SF6 gas and air circuit breakers at a relatively high current over several hundred amperes, during the "high current interval." It is used to investigate an attenuation occurrence as an influence of resistance of a circuit breaker arc.

It is expected that the series connection of two models is applicable for a wide range of currents. In this chapter, the tandem connection of the Mayr and Cassie arc models is used for analysis. In a high current region, the Cassie arc model is dominant as the arc resistance by the Mayr arc model is low. In a small current region, the Mayr arc model is dominant.

Power System Transient Analysis: Theory and Practice using Simulation Programs (ATP-EMTP), First Edition.
Eiichi Haginomori, Tadashi Koshiduka, Junichi Arai, and Hisatochi Ikeda.
© 2016 John Wiley & Sons, Ltd. Published 2016 by John Wiley & Sons, Ltd.
Companion website: www.wiley.com/go/haginomori_Ikeda/power

5.1   Mayr Arc Model

The arc conductance is expressed by the following differential equation by the Mayr arc model, assuming a constant arc diameter, a constant arc loss, and Saha's equation for conductivity:

$$\frac{1}{G}\frac{dG}{dt} = \frac{1}{\theta}\left(\frac{EI}{N_0} - 1\right), \tag{5.1}$$

where G = arc conductance, θ = arc time constant, E = arc voltage, I = arc current, and N_0 = constant arc loss.

Through introduction of Laplace operators, the equation can be transformed to

$$G_0 = I^2 / N_0 \tag{5.2}$$

$$G = G_0 / (1+\theta). \tag{5.3}$$

This equation can be expressed directly by the Transient Analysis of Control Systems (TACS) program of EMTP. The analyzed conductance G can be combined with an electric circuit and the calculation proceeds.

The Mayr arc model is suitable for a small current region below several tens of amperes. The model is just as suitable for the post-arc current region. While a fault current that a circuit breaker breaks is very large, the current breaking phenomenon is one near the current zero point, which is the reason why the Mayr arc model is suitable. Furthermore, the Mayr arc model is applicable to the arc, which is the region dominantly filled by so-called high-temperature gas. Therefore, the Mayr arc model is typically applied to the interaction interval. The period after the interaction interval is the high voltage interval, where the main phenomenon is dielectric and another model is applied.

5.1.1   Analysis of Phenomenon of Short-Line Fault Breaking

Figure 5.1 shows the basic circuit simplified from the circuit based on the International Electro technical Commission (IEC) standard for the short-line fault (SLF) breaking of 90% at the rated voltage 300 kV and the rated current 50 kA. The circuit breaker is simulated by the Mayr arc model. Here the following assumptions are introduced:

$$\theta\,(\text{time constant}) = 1\,\mu s$$
$$N_0\,(\text{arc loss}) = 293\text{ kW}.$$

The main part of the current breaking phenomenon is supposed to be a period of a few micro-seconds just before the current zero point. In addition, the distributed parameter line, which is extremely short, representing the initial transient recovery voltage (ITRV) is involved with the breaking phenomenon. In due course, a very short time step is required in an EMTP calculation. To reduce the total calculation time, TMAX, the initial condition of TACS variables should be chosen carefully.

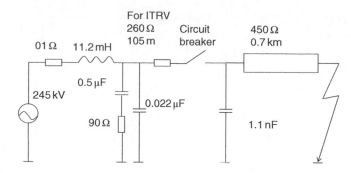

Figure 5.1   Short-line fault breaking circuit: 300 kV, 50 kA, L90 condition.

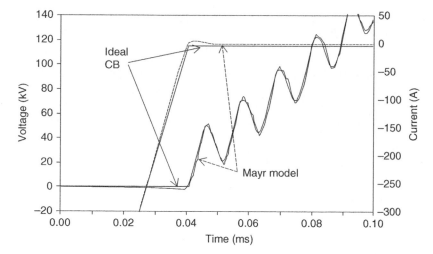

Figure 5.2   Calculation result for SLF by an ideal circuit breaker and circuit breaker with the Mayr arc model.

Figure 5.2 shows the calculated result. The breaking is carried out by both the ideal circuit breaker and the circuit breaker expressed by the Mayr arc model. The calculation time is 100 ms near the current zero point. The ideal circuit breaker is assumed to be without an arcing voltage and a rapid recovery of insulation after the current zero point. See the data files.[1,2]

The part near the current zero point is enlarged in Figure 5.3. The current zero point comes earlier with the circuit breaker of the Mayr arc model than that of the ideal circuit breaker. This is caused by the arcing voltage. The postarc current of several amperes flows in the reverse direction in the case of the Mayr arc model. The oscillation of the transient recovery voltage (TRV), which represents ITRV, is suppressed and the TRV becomes smooth in the case of the Mayr arc model as well.

[1] Data5-00: By ideal circuit-breaker, SLF breaking, 300 kV, 50 kA, L90 condition.
[2] Data5-01: Mayr arc model calculating SLF breaking, 300 kV, 50 kA, L90 condition.

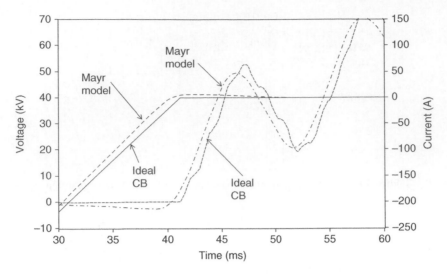

Figure 5.3   Enlargement around the current zero point of Figure 5.2.

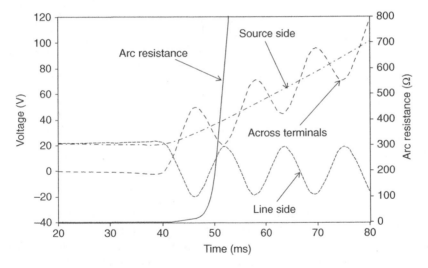

Figure 5.4   Arc resistance change near the current zero point.

Figure 5.4 shows the arc resistance changing along with time, together with the voltages of the source side, line side, and across the terminals of the circuit breaker. The arc resistance is low by the current zero point. The arc resistance then increases gradually after the current zero point and it rapidly increases toward an infinite value after the first peak value of the TRV. From the figure it can be understood that the ITRV oscillation is attenuated by the low arc resistance near the current zero point.

The results indicate the energy balance of arc after the current zero point. The comparison between the injected energy from the TRV and the energy loss from the arc decides the failure or success of the breaking. For the condition near the boundary of success or failure, the current starts to flow again at several microseconds after the current zero point when the arc time constant θ is made larger or the arc loss N_0 is made smaller.

Figure 5.5 shows the other testing circuit competent with the IEC standard, IEC62271-100, for the same condition as in Figure 5.1. The circuit is used for the purpose of simplifying the breaking test procedure. The component simulating ITRV is eliminated. On the other hand, the capacitance value of ramp constant is reduced from 1.1 to 0.22 nF at the line side of the circuit breaker.

By applying the following constant for the Mayr arc model,

$$\theta = 1.0\,\mu s$$
$$N_0 = 300\text{kW},$$

the calculation result shows that this condition is the boundary for the successful breaking. Figure 5.6 shows the TRVs and breaking currents calculated for both circuits of Figures 5.1

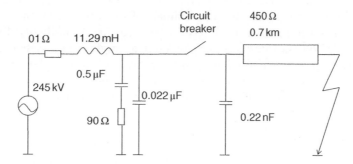

Figure 5.5   Alternative circuit for the condition in Figure 5.1.

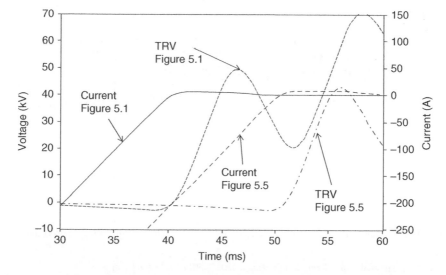

Figure 5.6   Breaking currents and TRVs of two circuits, Figures 5.1 and 5.5.

and 5.5. Both the postarc current and the TRV oscillation attenuation are different in comparison of two conditions. Both the postarc current and the TRV oscillation attenuation are bigger in the condition of Figure 5.5 than those in Figure 5.1. This phenomenon will differ by the choice of arc parameters. Further investigations are required. (See data file.[3])

5.1.2   Analysis of Phenomenon of Shunt Reactor Switching

When a shunt reactor magnetizing current is broken by a circuit breaker, the oscillation of current is generated during the period approaching the zero point. The relationship between the arcing current and the arcing voltage shows a negative characteristic, especially near the current zero point. In the condition of steady or quasi steady state, $dG/dt = 0$, $EI = N_0$ in the Mayr arc model equation. This means that the arcing voltage and current characteristic are negative. When the L-C circuit includes the negative resistance, the current oscillation appears expanded. The negative characteristics of arc were used for generating the oscillation at the early stage of radio technology.

Figure 5.7 shows the shunt reactor switching circuit of 300 kV and 150 MVA. The circuit is shown in the single-phase condition. The amplitude of shunt reactor magnetizing current is about several hundred amperes. A circuit breaker can break the current easily. The puffer type circuit breaker of SF6 gas, which is dominantly used in a high-voltage system, can break the current at the short arcing time when the distance between contacts is short and the blowing gas pressure is low. With this circuit breaker condition therefore, a low arc loss will be adequate in the Mayr arc model equation. Here, the arc time constant θ of 0.5 μs and the arc loss N_0 of 15 kW is assumed.

Figure 5.8 shows the calculation results. The current becomes zero by the oscillation that begins when the current is 8 A before the AC current zero point. The figure shows that the current is broken before the AC current reaches the normal current zero point. The arc time constant and arc loss chosen in the calculation are supposed to represent a modern gas circuit breaker characteristic, judging from the current value when the oscillation starts. (See data file.[4])

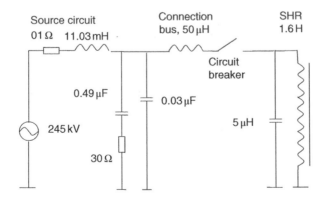

Figure 5.7   Shunt reactor switching circuit 300 kV, 150 MVA, single-phase representation.

---

[3] Data5-02: Mayr arc model calculating SLF breaking, 300 kV, 50 kA, L90 condition.
[4] Data5-03: Current chopping by breaking inductive current calculation—applying the Mayr arc model as circuit breaker.

The current appears to be chopped at 8A on a large scale. This phenomenon is called current chopping. The oscillation that occurs at time zero is generated intentionally for calculation by the circuit breaker operation, that is, the contact separation. This oscillation seems to supply a trigger of generating the oscillation near the current zero point.

Figure 5.9 shows the calculation result with an arc loss of twice as much as that for Figure 5.8. This arc loss comes from a condition of a long arcing time with a long contacts separation, along with a high blowing pressure or a series connection of plural contacts.

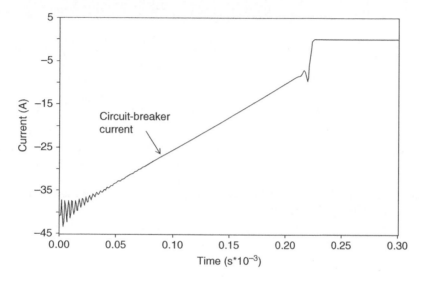

Figure 5.8   Current chopping by shunt reactor current breaking.

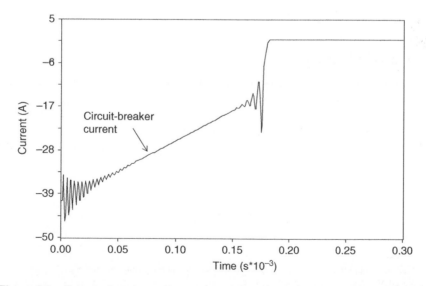

Figure 5.9   Current chopping at shunt reactor current breaking with doubled arc loss N_0.

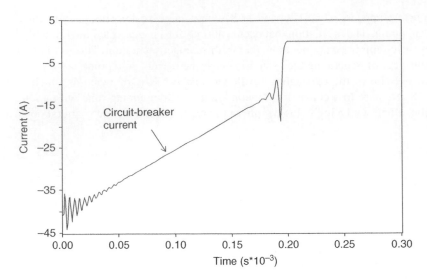

Figure 5.10   Same as Figure 5.8, but the reactor parallel capacitor is doubled from 5 to 10 nF.

The current value at which the oscillation starts becomes twice as much as that in Figure 5.8. The phenomenon follows the idea that the higher the quenching performance is, the larger the chopping current will be. It is known that the chopping current value is related to a route of the breaking point number. Another cause may have an effect by increasing the breaking point number. (See data file.[5])

Figure 5.10 shows the calculation result with the parallel capacitance of shunt reactor being twice as much as the one in Figure 5.8. The capacitances of 5 nF in a so-called second oscillation circuit are set to a double value. As a result, the chopping current value is related to the capacitance value as shown in references. (See data file.[6])

5.2   Cassie Arc Model

In the Cassie arc model, the arc is of a relatively high current (more than several hundred amperes) with the following assumptions:

1. Heat loss is dependent on arc flow and convection.
2. Heat loss, stored heat, and electrical conductance are proportional to the cross-sectional area expressed by the equation

$$\frac{1}{G}\frac{dG}{dt} = \frac{1}{\theta}\left(\frac{E^2}{E_0^2} - 1\right), \tag{5.4}$$

---

[5] Data5-04: Current chopping by a breaking inductive current calculation—applying the Mayr arc model as circuit-breaker, increased N_0.

[6] Data5-05: Current chopping by a breaking inductive current calculation—applying the Mayr arc model as circuit breaker.

where E = arc voltage, E_0 = constant, θ = arc time constant, and G = arc conductance. The Cassie arc model is applied to high current arcs.

By introducing the Laplace operator s, the equation can be transformed to

$$G_0 = G^2 \left(G = \sqrt{G_0} \right) \tag{5.5}$$

$$G = I / E \tag{5.6}$$

$$G_0 = \frac{I^2}{E_0^2} \frac{1}{1+\theta s} \tag{5.7}$$

$$R = 1 / G. \tag{5.8}$$

This equation can be expressed directly by the TACS program of EMTP, as with the Mayr arc model. The analyzed conductance G can be combined with an electric circuit and the calculation proceeds.

At the steady-state condition, dG/dt is zero and then the arc voltage E is equal to E_0. It is important that an adequate value of E_0 is selected.

5.2.1   Analysis of Phenomenon of Current Zero Skipping

In AC circuits, the current zero appears at every half cycle. When the DC component is large enough, the current skipping phenomenon without the current zero appears.

The arc voltage of circuit breaker produces the current zero to the zero skipping current. Figure 5.11 shows the circuit in which the effect of arc voltage is analyzed when the circuit breaker is open in the circuit with zero skipping current generation.

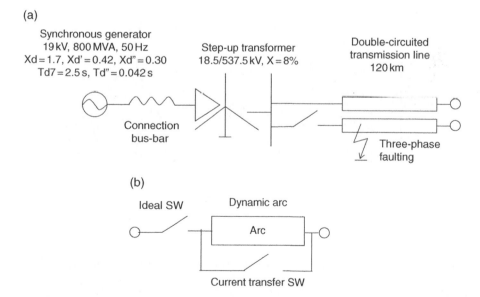

(a)

Synchronous generator
19 kV, 800 MVA, 50 Hz
Xd = 1.7, Xd' = 0.42, Xd" = 0.30
Td7 = 2.5 s, Td" = 0.042 s

Step-up transformer
18.5/537.5 kV, X = 8%

Double-circuited
transmission line
120 km

Connection
bus-bar

Three-phase
faulting

(b)

Ideal SW   Dynamic arc

Arc

Current transfer SW

Figure 5.11   Circuit for zero skipping current breaking calculation. (a) Power system diagram. (b) Circuit diagram with arc dynamics.

Figure 5.11a shows the circuit where the generator supplies the charging current of transmission line through the step-up transformer. The earth faults occur at three phases of two lines near the 550 kV substation bus bar. The earth faults do not occur simultaneously but separately at the phases where the zero skipping current goes to maximum value. In this calculation, the earth fault at A, B, and C phases occurs at the times of 14, 8.3, and 8.3 ms, respectively, when the time zero is settled at the peak of the phase A voltage. The condition is obtained by a cut-and-try repetition and will hardly occur in a real system. The important parameters of generator are shown in Figure 5.11a.

Figure 5.11b shows the circuit by which the dynamic arc characteristic of circuit breaker is calculated by the TACS program. The TACS program does not operate in the condition of $t < 0$, and the switch should be located in parallel with it.

Figure 5.12 shows the calculated three-phase currents. The current zero does not occur in phases A and C. The condition might not happen in a real system statistically, but it is selected to investigate the phenomenon. (See data file.[7])

Figure 5.13 shows the calculation result with an ideal circuit breaker that has no arc voltage and breaks the current at the first current zero point after the operation signal is given. The operation signal is given at the time of 48 ms. Current zero comes at phase B, in which the current is broken at the first current zero point after the signal is given. As phase B is broken, the circuit becomes asymmetrical, which results in making the zero phase current. As a result, the DC component in the phase A and C currents attenuate rapidly and the current zeros are produced. Two phases are broken consequently. (See data file.[8])

Figure 5.14 shows the calculation result of zero skipping current when the arc voltage is simulated by the Cassie arc model with $E_0 = 1000 \text{V}$ for the circuit breaker. The result in

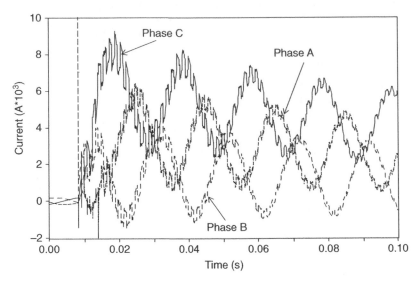

Figure 5.12 Zero skipping short-circuit current without influence of circuit-breaker arc.

---

[7] Data5-11: Zero skipping current breaking near generator—fault current lasting.
[8] Data5-12: Zero skipping current breaking near generator—fault current interruption by ideal SW—zero arc voltage.

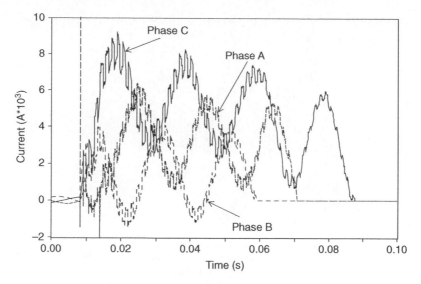

Figure 5.13   Breaking by ideal circuit breaker.

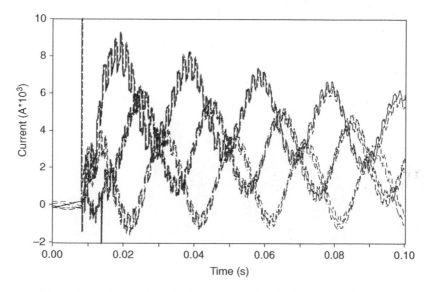

Figure 5.14   Three-phase fault currents with and without arc voltages.

Figure 5.12 without the arc voltage is included in the figure for comparison. The difference between the two cases is small. It is understood that the reason is due to the low arc voltage compared with the system voltage and a small contribution to the current attenuation. (See data file.[9])

Figure 5.15 shows the calculation result in which the current is broken by the circuit breaker of the Cassie arc model. As the comparison shows in Figure 5.14, the zero skipping current is

---

[9] Data5-13: Zero skipping current breaking near generator—dynamic arc introduced, still nonbreaking.

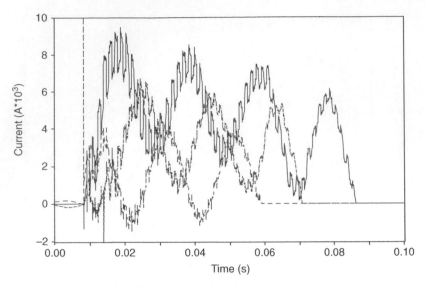

Figure 5.15   Fault current interruption by a circuit breaker with the Cassie arc model: $E_0 = 1000\,\text{V}$.

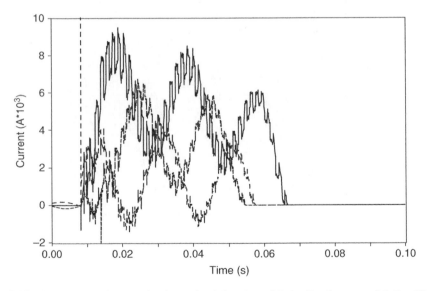

Figure 5.16   Fault current interruption by a circuit breaker with the Cassie arc model: $E_0 = 10,000\,\text{V}$.

hardly influenced by the arc voltage of $E_0 = 1000\,\text{V}$, and the current breaking hardly changes. It is understood that the arc voltage has negligible influence on the zero skipping current breaking in a 550 kV system.

The arc voltage with a Cassie arc model of $E_0 = 10,000\,\text{V}$ is assumed, although it is an unrealistic one. The calculation result with the condition is shown in Figure 5.16. Due to the high arc voltage, the current zero point comes earlier in the phase where the current zero skipping occurs. As a result, the fault period is reduced. (See data file.[10])

The following points are a summary of the zero skipping current breaking phenomenon in a 500 kV system.

- The serious zero skipping current depends on the timing of the fault occurrence and occurs rarely.
- On the occasion of such a special case, the current of one phase in three has a current of zero. The current in the phase can be broken by a usual circuit breaker. When one phase current is broken, the zero sequence resistance becomes included in the circuit. As the DC component is attenuated by the resistance, the current zero is derived in the other phases. As a result, a usual circuit breaker can break it.
- In a 500 kV system, the arc voltage of the Cassie arc model with $E_0 = 10,000$ V has an effect of making the current zero come earlier in the zero skipping current.

In the case of a two-phase fault, the zero sequence is not related and the phenomenon becomes much more complicated. The AC component of fault current is not attenuated much in this case, and the current zero comes early. The calculation result is shown in Figure 5.17. (See data file.[11])

In this chapter, the zero skipping current was discussed pertaining to a high-voltage system with a high circuit breaker. The phenomena associated with the superposition of the capacitive current on the shunt reactor current or the generator circuit itself should be investigated, respectively.

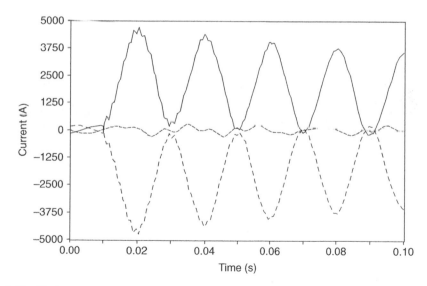

Figure 5.17 Fault current in the case of a two-phase fault by a circuit breaker with the Cassie arc model: $E_0 = 1000$ V.

---

[10] Data5-14: Zero skipping current breaking the near generator—fault current interruption by the Cassie arc + ideal SW.
[11] Data5-15: Zero skipping current breaking the near generator—fault current lasting. Two-phase isolated faulting.

Appendix 5.A: Mayr Arc Model Calculating SLF Breaking, 300 kV, 50 kA, L90 Condition

The circuit diagrams in the data file[2] are shown in Figure A5.1.

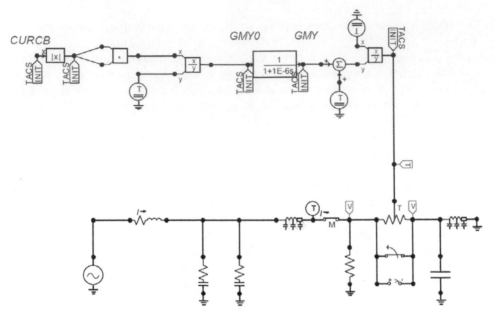

Figure A5.1   Circuit diagram in data file.[2]

The data for the circuit components are shown in Figures A5.2–A5.12.

Figure A5.2   Circuit components data (1).

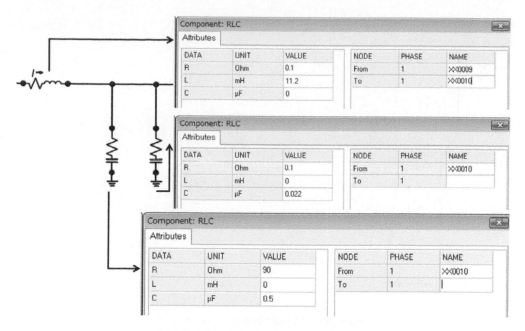

Figure A5.3   Circuit components data (2).

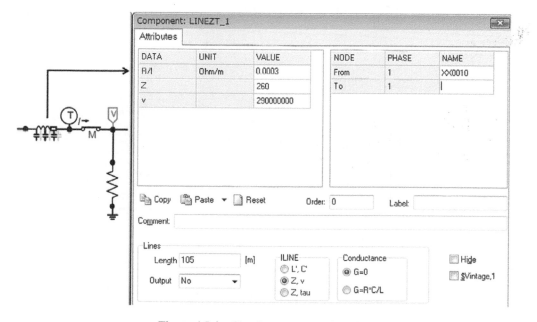

Figure A5.4   Circuit components data (3).

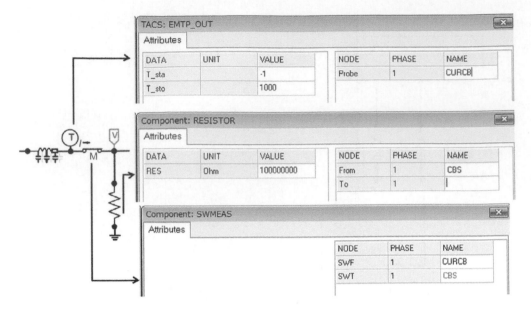

Figure A5.5   Circuit components data (4).

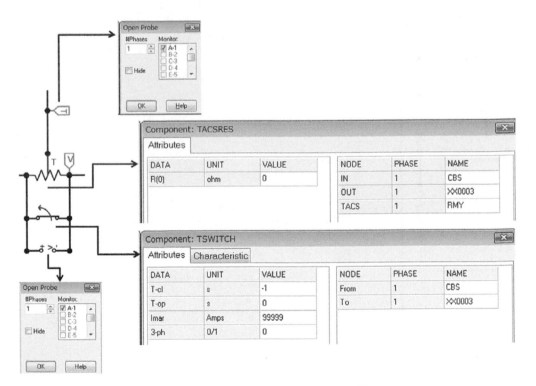

Figure A5.6   Circuit components data (5).

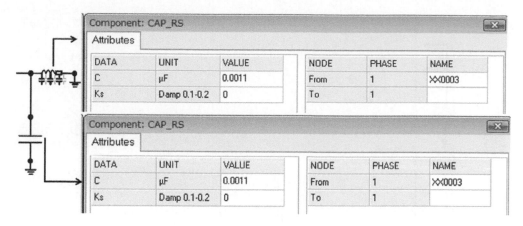

Figure A5.7   Circuit components data (6).

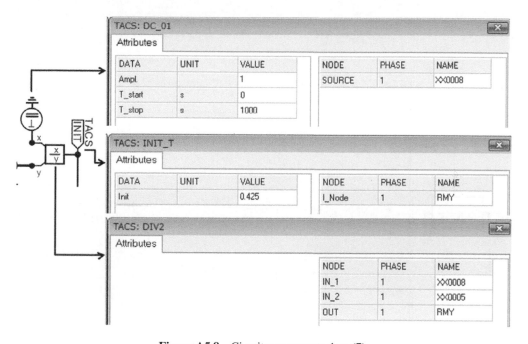

Figure A5.8   Circuit components data (7).

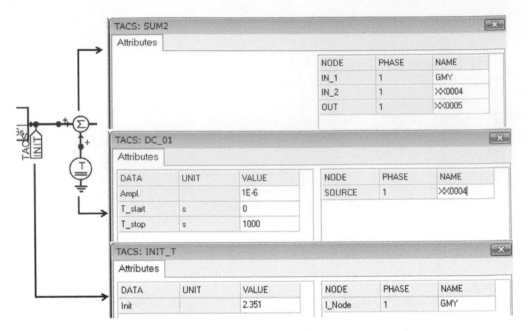

Figure A5.9   Circuit components data (8).

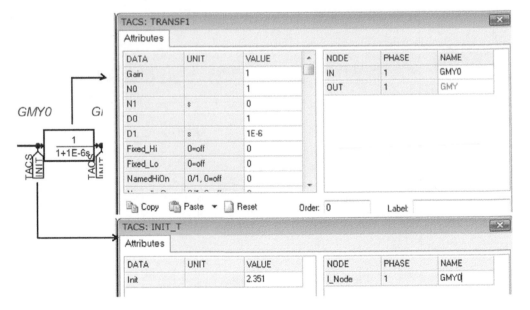

Figure A5.10   Circuit components data (9).

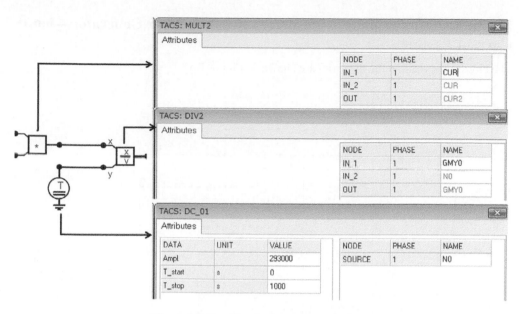

Figure A5.11   Circuit components data (10).

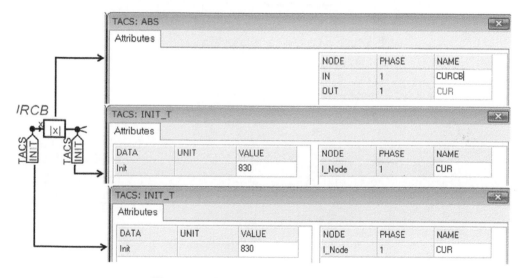

Figure A5.12   Circuit components data (11).

Appendix 5.B: Zero Skipping Current Breaking Near Generator—Fault Current Lasting

The circuit diagrams in the data file[7] are shown in Figure B5.1.

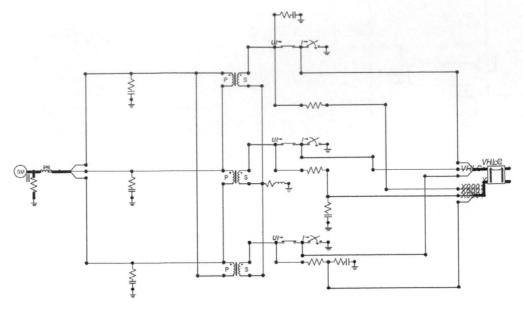

Figure B5.1   Circuit diagram in data file.[7]

The data for the circuit components are shown in Figures B5.2–B5.12.

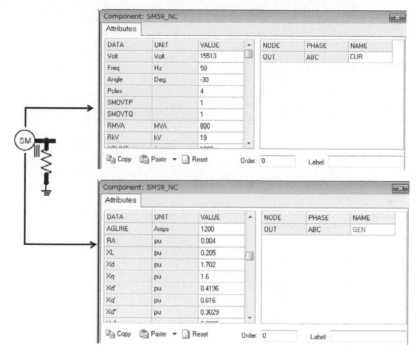

Figure B5.2   Circuit components data (1).

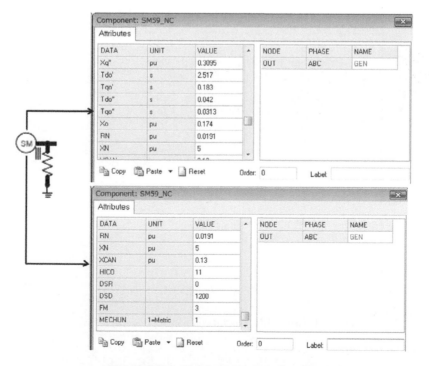

Figure B5.3   Circuit components data (2).

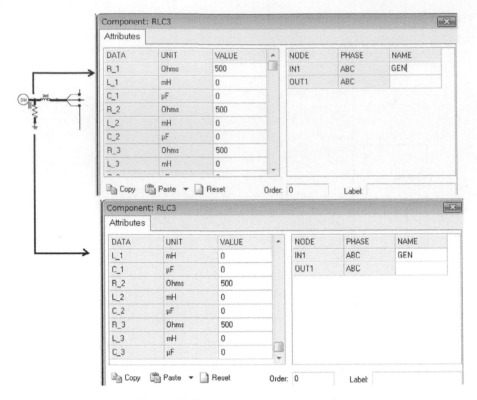

Figure B5.4   Circuit components data (3).

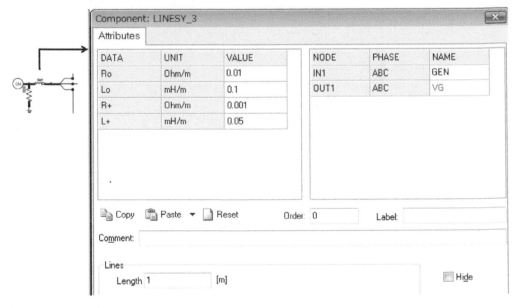

Figure B5.5   Circuit components data (4).

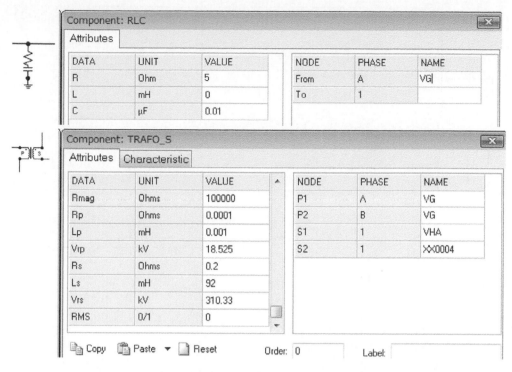

Figure B5.6 Circuit components data (5).

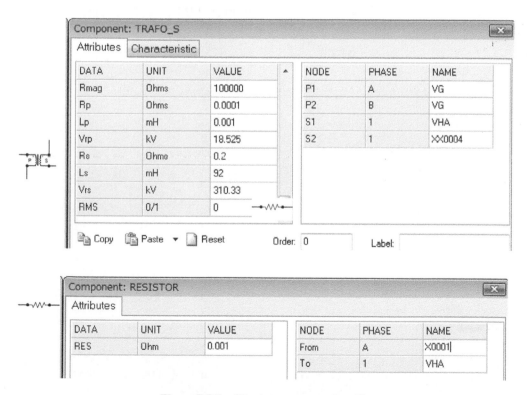

Figure B5.7 Circuit components data (6).

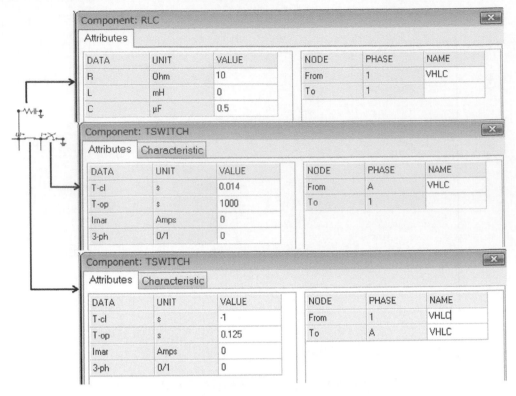

Component: RLC

Attributes

DATA	UNIT	VALUE		NODE	PHASE	NAME
R	Ohm	10		From	1	VHLC
L	mH	0		To	1	
C	µF	0.5				

Component: TSWITCH

Attributes   Characteristic

DATA	UNIT	VALUE		NODE	PHASE	NAME
T-cl	s	0.014		From	A	VHLC
T-op	s	1000		To	1	
Imar	Amps	0				
3-ph	0/1	0				

Component: TSWITCH

Attributes   Characteristic

DATA	UNIT	VALUE		NODE	PHASE	NAME
T-cl	s	-1		From	1	VHLC
T-op	s	0.125		To	A	VHLC
Imar	Amps	0				
3-ph	0/1	0				

Figure B5.8   Circuit components data (7).

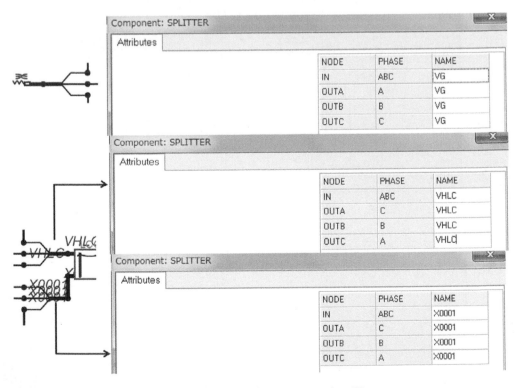

Component: SPLITTER

Attributes

NODE	PHASE	NAME
IN	ABC	VG
OUTA	A	VG
OUTB	B	VG
OUTC	C	VG

Component: SPLITTER

Attributes

NODE	PHASE	NAME
IN	ABC	VHLC
OUTA	C	VHLC
OUTB	B	VHLC
OUTC	A	VHLC

Component: SPLITTER

Attributes

NODE	PHASE	NAME
IN	ABC	X0001
OUTA	C	X0001
OUTB	B	X0001
OUTC	A	X0001

Figure B5.9   Circuit components data (8).

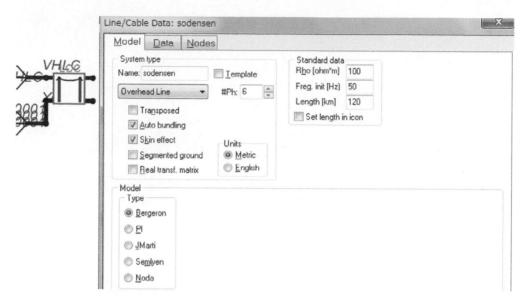

Figure B5.10   Circuit components data (9).

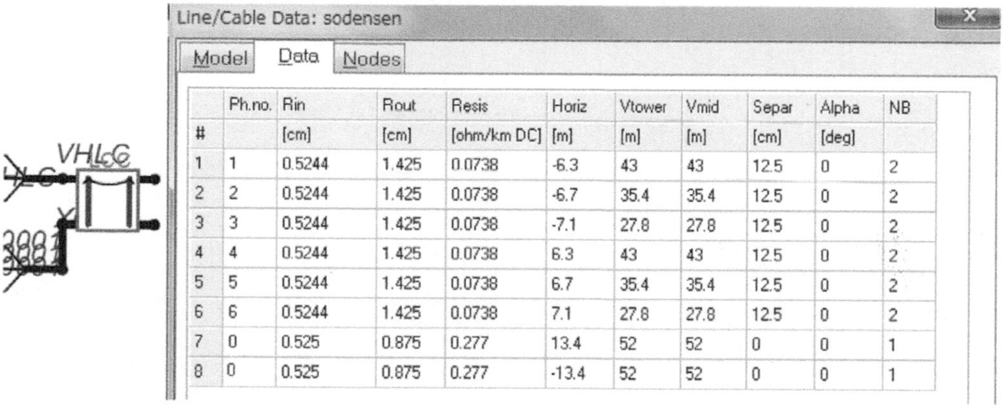

#	Ph.no.	Rin [cm]	Rout [cm]	Resis [ohm/km DC]	Horiz [m]	Vtower [m]	Vmid [m]	Separ [cm]	Alpha [deg]	NB
1	1	0.5244	1.425	0.0738	-6.3	43	43	12.5	0	2
2	2	0.5244	1.425	0.0738	-6.7	35.4	35.4	12.5	0	2
3	3	0.5244	1.425	0.0738	-7.1	27.8	27.8	12.5	0	2
4	4	0.5244	1.425	0.0738	6.3	43	43	12.5	0	2
5	5	0.5244	1.425	0.0738	6.7	35.4	35.4	12.5	0	2
6	6	0.5244	1.425	0.0738	7.1	27.8	27.8	12.5	0	2
7	0	0.525	0.875	0.277	13.4	52	52	0	0	1
8	0	0.525	0.875	0.277	-13.4	52	52	0	0	1

Figure B5.11   Circuit components data (10).

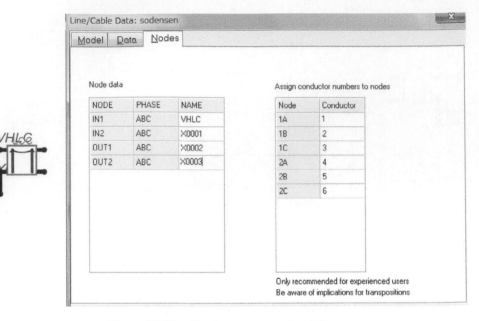

Figure B5.12   Circuit components data (11).

Appendix 5.C: Zero Skipping Current Breaking Near Generator— Dynamic Arc Introduced, Still Nonbreaking

The circuit diagrams in the data file[9] are shown in Figure C5.1.

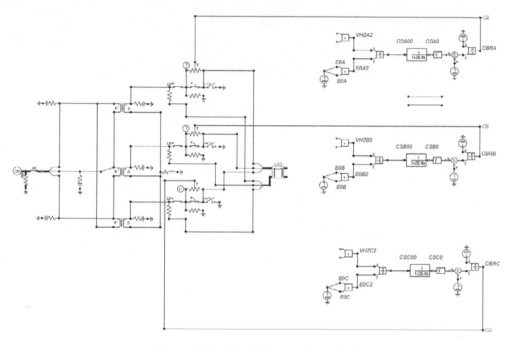

Figure C5.1 Circuit diagram in data file.[9]

The data for the circuit components are shown in Figures C5.2–C5.7.

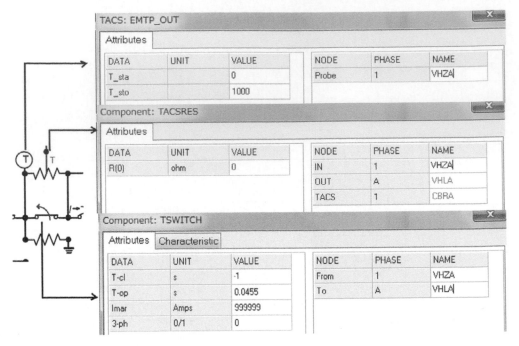

Figure C5.2   Circuit components data (1).

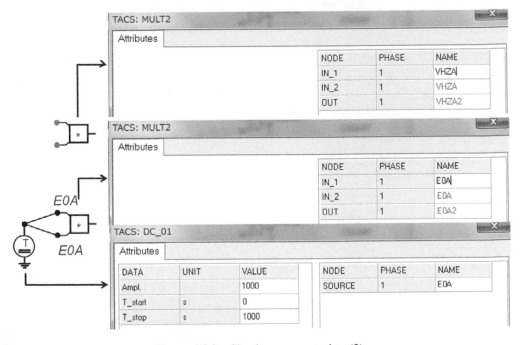

Figure C5.3   Circuit components data (2).

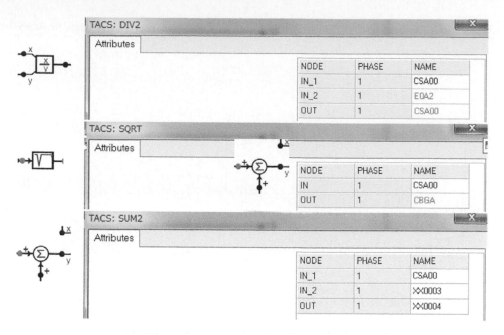

Figure C5.4 shows TACS: DIV2, TACS: SQRT, and TACS: SUM2 windows.

TACS: DIV2 — Attributes

NODE	PHASE	NAME
IN_1	1	CSA00
IN_2	1	E0A2
OUT	1	CSA00

TACS: SQRT — Attributes

NODE	PHASE	NAME
IN	1	CSA00
OUT	1	CBGA

TACS: SUM2 — Attributes

NODE	PHASE	NAME
IN_1	1	CSA00
IN_2	1	XX0003
OUT	1	XX0004

Figure C5.4 Circuit components data (3).

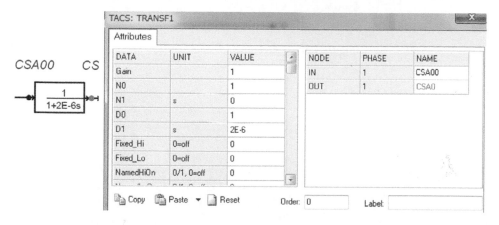

TACS: TRANSF1 — Attributes

CSA00   CS

$$\frac{1}{1+2E\text{-}6s}$$

DATA	UNIT	VALUE		NODE	PHASE	NAME
Gain		1		IN	1	CSA00
N0		1		OUT	1	CSA0
N1	s	0				
D0		1				
D1	s	2E-6				
Fixed_Hi	0=off	0				
Fixed_Lo	0=off	0				
NamedHiOn	0/1, 0=off	0				

Copy Paste ▼ Reset   Order: 0   Label:

Figure C5.5 Circuit components data (4).

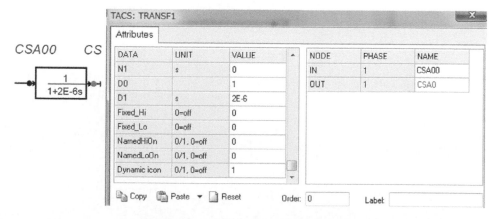

TACS: TRANSF1 — Attributes

CSA00   CS

$$\frac{1}{1+2E\text{-}6s}$$

DATA	UNIT	VALUE		NODE	PHASE	NAME
N1	s	0		IN	1	CSA00
D0		1		OUT	1	CSA0
D1	s	2E-6				
Fixed_Hi	0=off	0				
Fixed_Lo	0=off	0				
NamedHiOn	0/1, 0=off	0				
NamedLoOn	0/1, 0=off	0				
Dynamic icon	0/1, 0=off	1				

Copy Paste ▼ Reset   Order: 0   Label:

Figure C5.6 Circuit components data (5).

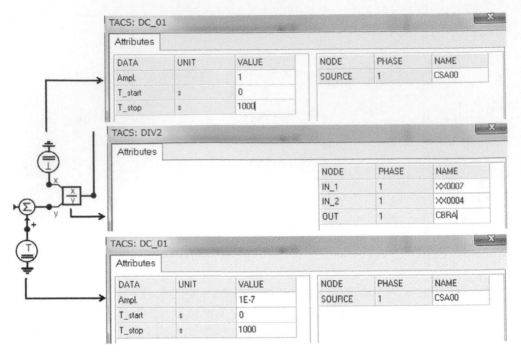

Figure C5.7   Circuit components data (6).

6

Typical Power Electronics Circuits in Power Systems

6.1 General

More and more widely, power electronics technologies/facilities are being applied to power systems [1]. In this chapter, several typical and primitive circuits are surveyed regarding their principal functions. Please note that basic studies of power electronics facilities in general have to be made beforehand [1]. This chapter mainly concerns practical applications.

6.2 HVDC Converter/Inverter Circuits

In HVDC (high-voltage DC transmission) systems for the purpose of AC-DC converting/inverting, thyristor (externally communicated) type facilities are generally applied due to, mainly, economical reasons. Figure 6.1a,b shows basic HVDC converter and inverter circuits, respectively.[1,2] In the figures only positive-polarity-side circuits are shown. In actual systems, negative-polarity ones, the circuits of which are the same as the positive ones, also exist where the DC upper side is the earth and the lower side is the negative polarity of the DC transmission line. By such a transmission circuit arrangement, the ground (return) current is minimized.

In Figure 6.1, some supplemental elements, such as snubbers, capacitances around transformers, and so on, are not shown. For details, see the Electromagnetic Transients Program (EMTP) data files.[1] In the relevant actual circuits, some further additional (supplemental) elements might be necessitated.

First, the performance of the converter is surveyed. Connecting 50 Hz three-phase source circuits together with the source impedances to the left-hand side of Figure 6.1a, and the load

---

[1] ATPData6-01.dat/Data6-1.acp: HVDC transmission converter circuit, 50 Hz, 275 kV to DC, 250 kV, alpha (α) = 18°, ref. Figure 6.1a.
[2] ATPData6-02.dat/Data6-2.acp: HVDC transmission inverter circuit, DC, 250 kV to 60 Hz, 275 kV, beta (β) = 120°, ref. Figure 6.1b.

Power System Transient Analysis: Theory and Practice using Simulation Programs (ATP-EMTP), First Edition.
Eiichi Haginomori, Tadashi Koshiduka, Junichi Arai, and Hisatochi Ikeda.
© 2016 John Wiley & Sons, Ltd. Published 2016 by John Wiley & Sons, Ltd.
Companion website: www.wiley.com/go/haginomori_Ikeda/power

(a)

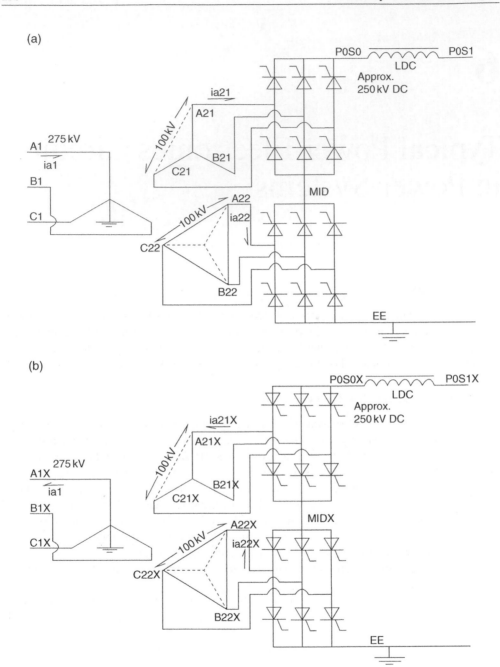

(b)

Figure 6.1   High-voltage DC transmission (HVDC) 275 kV 50 Hz–DC–275 kV 60 Hz system circuits.[1,2] (a) HVDC converting circuit.[1] (b) HVDC inverting circuit.[2]

resistor to the right-hand side, calculation is made. The key point is the timing of the thyristor gate ignition signals. In the data file, the Transient Analysis of Control Systems (TACS) function is applied for this purpose. Typical results are shown in Figure 6.2.

As shown in Figure 6.1a, two of the valve bridge-sets (six-armed each), connected to the transformer star and delta windings, work for converting. Each bridge produces DC voltage with ripples of 60 electrical degrees of period. The sum of these two DC outputs is the total DC output voltage. Due to the phase voltage angle staggering in delta/star transformer windings of 30°, the ripple in DC voltage is well mitigated. At the same time, harmonics in the AC side current are diminished. Figure 6.2a–d shows such an effect. Each upper or lower side

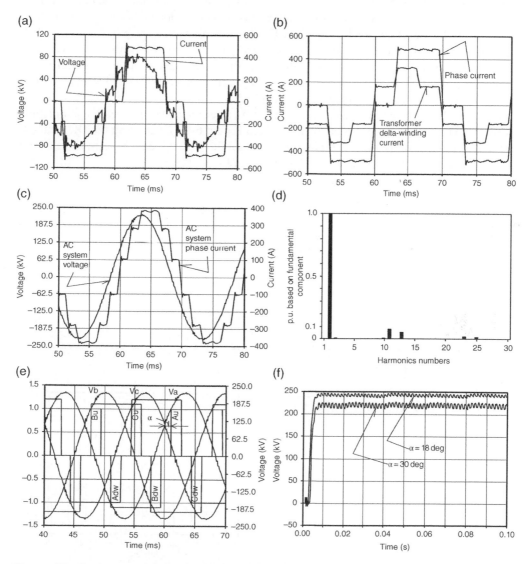

Figure 6.2 Performances of HVDC converter.[1] (a) Star winding voltage and current. (b) Currents in delta/phase. (c) AC system side voltage and current. (d) Fourier spectrum of the current. (e) AC voltage and gate signals. (f) α (delay time) versus DC voltage.

bridge phase current is square shaped (mainly by the effect of the DC circuit series reactor). But combining both the star and delta windings' currents in the transformer, the AC side current is well sinusoidally shaped. The effect is clearly shown in the Fourier spectrum (d), where only the low value $[12n \pm 1]$ order harmonics exist. The filter's capacitance for eliminating such high frequency harmonics components can be smaller compared to the lower frequency harmonics case.

Valve gate signal timing is to be based on the relevant AC voltage wave shape (phase angle) to be applied to the bridge, which is shown in Figure 6.2e. In this case, so-called ignition delay angle (alpha or α) is 18°. In Figure 6.2f, output DC voltage is shown, together with the case of ignition delay (alpha) equal to 30°. Theoretically, by primitive estimation, the output voltage is proportional to the cosine of the delay angle ($\cos\alpha$), that is, alpha=0 corresponds to diode function. To further eliminate the ripple in DC, a higher value of DC reactor can be inserted.

Next, let's study the externally communicated inverter's performance. For very high capacity systems, such as for a power utility's HVDC ones, due to an economical and/or efficiency point of view, externally communicated systems are mostly applied where relatively cheap and high capacity of thyristors can be applied.

Figure 6.1b shows an inverter circuit, where (for easier understanding) only the directions of the power flow and thyristors are reversed from Figure 6.1a. Connecting the DC source to the right-hand side of the inverter bridge, and 60 Hz of the AC source to the left-hand side, the performance is analyzed.[2] Some results are shown in Figure 6.3.

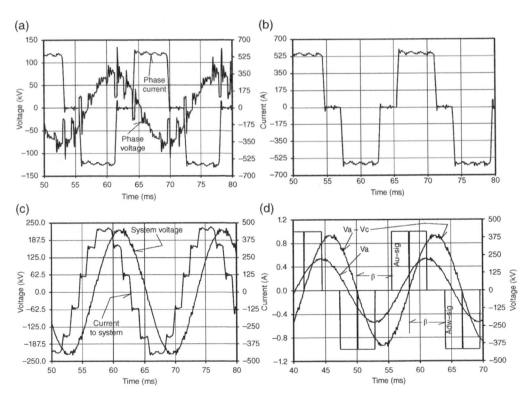

Figure 6.3 HVDC inverter performances.[2] (a) Star winding voltage and current. (b) Delta side phase current. (c) AC side voltage and current. (d) Voltage and gate signals.

It should be noted that the gate signal timing (of each pole of the thyristor) is to be further delayed compared to the converter of case (>90°) from the zero point of the applied voltage (i.e., Va–Vc for the pole of "a"), as shown in (d), thus resulting in a normal externally communicated inverter performance. The delay angle is usually called beta or β. Currents in both the upper and lower side valves are square wave shaped ones like in the converter, and by combining these in the AC side via the delta–wye winding connections in the transformer, the well-formed AC current (60 Hz) is produced. The AC current value is controlled by both DC voltage (by alpha in converter) and advanced angle (beta). Note that for the current supplied to the AC system, capacitive reactive current (due to beta) and some harmonics components are involved. For absorbing these, a capacitor bank combined with filters is installed.

Finally, to survey HVDC system performance, the right-hand side of the converter in Figure 6.1a (DC output) is connected to the right-hand side of the inverter in Figure 6.1b via a DC transmission line represented by a reactor and resistor, while excluding a DC source and DC load resistor. This establishes a single-pole HVDC transmission system, transferring power from 50 to 60 Hz AC via DC transmission.[3,4] The calculation results at the start up of the system are shown in Figure 6.4. A lower alpha value corresponds to a higher DC voltage and

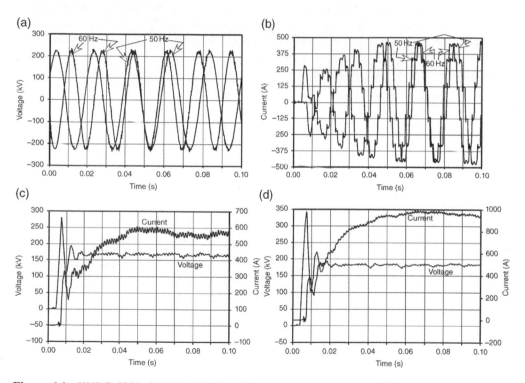

Figure 6.4 HVDC 50 Hz 275 kV–DC–60 Hz 275 kV transmission system.[3,4] (a) 50 Hz/60 Hz system voltages. (b) 50 Hz/60 Hz system currents. (c) DC voltage and current ($\alpha = 45°$).[3] (d) DC voltage and current ($\alpha = 35°$).[4]

---

[3] ATPData6-03.dat/Data6-3.acp: HVDC transmission circuit, 50 Hz, 275 kV (DC +250/−0 kV) 60 Hz, 275 kV, $\alpha = 45°$, $\beta = 120°$, approximately 100 MVA transmitting.

[4] ATPData6-04.dat: Same as data file 3, but $\alpha = 35°$, $\beta = 120°$, approximately 180 MVA transmitting.

transmission power, as shown in Figure 6.4c ($\alpha=45°$) and 6.4d ($\alpha=35°$). In actual systems, highly accurate gate ignition control seems to be especially important.

6.3   Static Var Compensator/Thyristor-Controlled Inductor

For compensation purpose by controlling reactive power quickly in power systems, most generally and widely thyristor controlled inductor type Static Var Compensators (SVCs) are applied. The facility itself can control only inductive reactive power. Therefore, for controlling capacitive reactive power as well, the capacitor bank must be connected in parallel.

Figure 6.5 shows a basic three-phase SVC circuit, where some additional elements, such as snubbers, stray capacitances, and so on, are not shown.[5–8] For the sake of the three-phase circuit continuity, since the reactor current is not continuous through the period of power frequency (Figure 6.6a), thyristor-controlled inductors are connected between phases (delta connection). Detailed circuit constants are shown in the attached data files.[5–8] By controlling the time interval of current flowing (current loop duration) by the thyristor in each cycle, the inductive reactive power is equivalently controlled. A narrow current loop width corresponds to the lower reactive power. The current wave shape is neither sinusoidal nor continuous.

Here, a three-phase SVC rated 6.6 kV, 3000 kVar (at maximum) is taken up. Some calculated results are shown in Figure 6.6. SVC controlling is based on α—the ignition delay angle, by which the current flowing duration is controlled—together with the crest value. By a rough estimation calculation, the crest value and the loop width of the current are proportional to

$$(1-\sin\alpha) \text{ and } (90-\alpha)/90,$$

respectively. The calculated results in Figure 6.6 give the following:

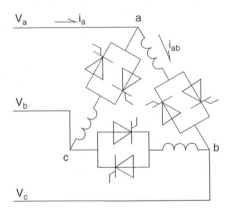

Figure 6.5   Thyristor-controlled reactor circuit.[5–8]

[5] ATPData6-11.dat/Data6-11.acp: Three-phase thyristor-controlled reactor (SVC), 6.6 kV, 1000 kVA, $\alpha=30°$, refer to Figures 6.5 and 6.6.
[6] ATPData6-12.dat: Same as data file 5, but 500 kVA, $\alpha=45°$, refer to Figures 6.5 and 6.6.
[7] ATPData6-13.dat: Same as data file 5, but 200 kVA, $\alpha=55°$, refer to Figures 6.5 and 6.6.
[8] ATPData6-14.dat: Same as data file 5, but 3000 kVA (Rated), $\alpha=5°$, refer to Figures 6.5 and 6.6.

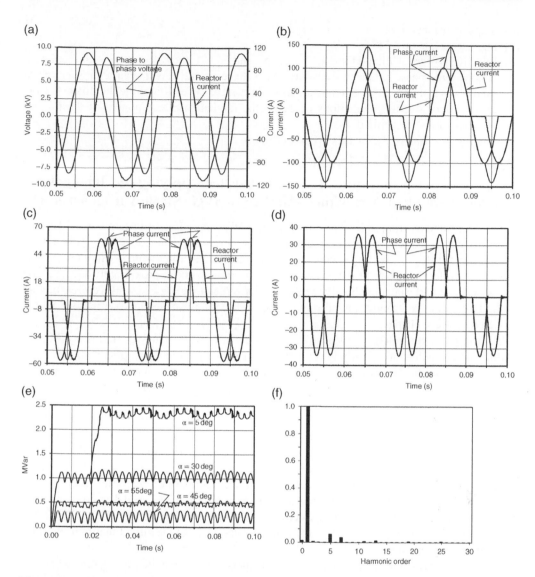

Figure 6.6 Thyristor-controlled reactor (SVC) performance.[5-8] (a) Phase-to-phase voltage and reactor current.[5] (b) Reactor currents and phase current.[5] (c) Same as (b) but $\alpha=45°$.[6] (d) Same as (b) but $\alpha=55°$.[7] (e) α versus reactive power.[5-8] (f) Fourier spectrum of (b) current.[5]

(a) Phase-to-phase voltage and current in the inductor connected between phases with $\alpha=30°$. The current is neither sinusoidal nor continuous.[5]

(b) Phase current at the top of the delta and within-delta reactor current ($\alpha=30°$). The phase current wave shape is a well-formed sinusoid. See the Fourier spectrum shown in (f).[5]

(c) Same as (b), but $\alpha = 45°$.[6]
(d) Same as (b), but $\alpha = 55°$.[7]
(e) Reactive power calculated for $\alpha = 5°$, $30°$, $45°$, and $55°$.[5-8] The calculation basis (equation) is for a three-phase balanced sinusoidal wave shape, so the actual accurate power values might be a little different. See the equations in the TACS section of the data files. For the most correct values, calculations based on the fundamental component of the Fourier spectrum are to be applied.
(f) Fourier spectrum of the phase current in (b).[5]

6.4 PWM Self-Communicated Type Inverter Applying the Triangular Carrier Wave Shape Principle—Applied to SVG (Static Var Generator)

The basic principle of a pulse width modulated (PWM) inverter is the same as a DC step-down chopper, where a constant frequency of square wave–shaped voltage pulses with constant crest value, the width of which is proportional to the target voltage (duty ratio). This produces current in the circuit with a certain value of inductance approximately equal to 1 by the DC voltage of the target value. By a relatively slower change of pulse width corresponding to the target voltage change, the current change will be almost the same as the one by the target DC voltage change.

By using a target (reference) sinusoidal voltage wave shape of power frequency AC, the frequency of which is sufficiently lower than the pulse voltage frequency, practically the same current to that of the AC source voltage is produced. As the most simple/primitive method to obtain the appropriate pulse width, the method of comparing a triangular-shaped carrier wave and the target (reference) AC wave shape is often applied.

The principle circuit diagram is shown in Figure 6.7.[9] Some additional elements such as snubbers, stray capacitances, and so on, are not shown in the figure. For details, refer to the data files. The neutral point of the load circuit is to be earthed through high impedance (to be explained later). When bipolar switching elements are applied, the circuit around the switching elements can be as shown in the figure. In the case of monopolar switching elements, such as gate turn-off thyristors (GTOs), diodes are to be connected in parallel to the switching elements. Please refer to other power electronics textbooks [1] for details. In EMTP, the No. 13 switching element, here applied, is an ideal bipolar switch, so the circuit diagram as shown in the figure is applicable as it is.

The control principle to produce the voltage pulses is shown in Figure 6.8, where by comparing the triangular wave shape to the reference voltage wave shape, an appropriate pulse width (duty ratio) corresponding to the phase-to-phase reference voltage value can be produced. Please note, in Figure 6.8, time P2–P3, P4–P5, or P6–P7 is proportional to ref A–ref B values at the times, corresponding to each phase-to-phase voltage.

Care should be taken that, in Figure 6.7, appropriate phase-to-phase voltage to the load circuit is applied, but the voltage at the neutral point (at NN) fluctuates much. Therefore, the neutral point is never solidly earthed. If a solidly (or by low impedance) earthed neutral load circuit is required, another circuit diagram is to be applied.

---

[9] ATPData6-21.dat/Data6-21.acp: Three-phase PWM inverter, basic/most simplified circuit, refer to Figures 6.7–6.9.

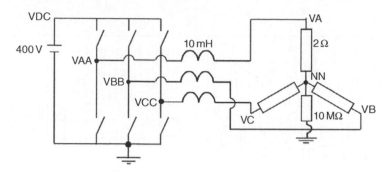

Figure 6.7   Self-communicated inverter circuit.[9]

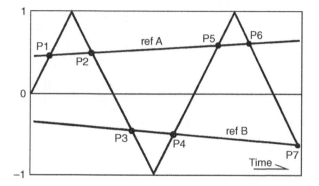

Figure 6.8   Triangular carrier wave versus reference voltage wave.

Some calculated results[9] are shown in Figure 6.9:

(a) Control commands in TACS, by which gate signals to switch elements in the inverter circuit are produced (for switch-ON state).
(b) Actually applied voltage wave shape between phase A and phase B. At approximately 40 ms, the control signal VA0–VB0 in (a) is maximum. At that timing in (b), pulse width is maximum; thus, the pulse width is made proportional to the instantaneous crest value of the target (reference) voltage.
(c) Fourier spectrum of the voltage in (b), where harmonics of the triangular carrier wave frequency and its integral numbers are significant. As inductively reactive components (i.e., inductance) are involved in the load circuit, high-frequency harmonics in the load current are not significant as shown in (d) and (e).
(d) Current in the load circuit. Wave shape distortion level is very low.
(e) Same current as (d) on the Fourier spectrum. Harmonics component is negligible.
(f) Introducing probable DC source impedance[10] due to the current chopping by the switching elements, overvoltages are created (ATPData6-22.dat).[10] Introducing snubbers in parallel to the switching elements (ATPData6-23.dat), the overvoltages are mitigated.[11]

---

[10] ATPData6-22.dat: Same as data file 9, but with DC source impedance elements involved; refer to Figures 6.7–6.9.
[11] ATPData6-23.dat: Same as data file 9, but snubbers are connected; Figures 6.7–6.9.

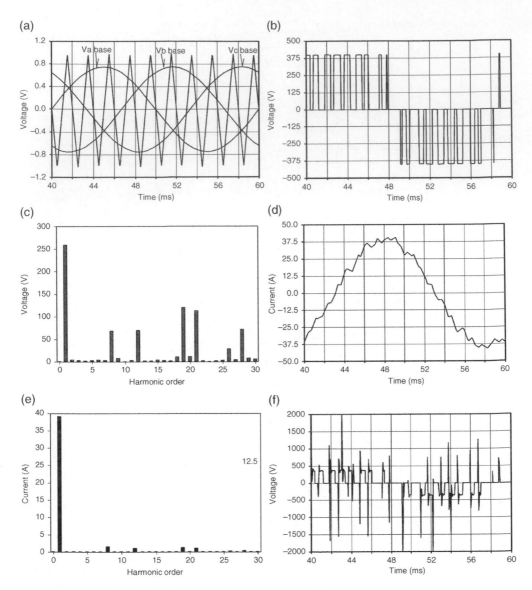

Figure 6.9   Self-communicated PWM inverter performance.[9,10] (a) Triangular carrier and reference voltages. (b) Phase-to-phase applied voltage wave. (c) Fourier spectrum of the wave in (b). (d) Load circuit current. (e) Fourier spectrum of the current (d). (f) Realistic circuit creating PWM voltage.[10]

As an application of a PWM self-communicated-type inverter, a static var generator (SVG) is shown in Figure 6.10, which is the most simplified circuit diagram.[12] A three-phase PWM inverter, which, seen from the load side, is equivalent to a three-phase AC voltage source, is

───────────────

[12] ATPData6-25.dat/Data6-25.acp: Three-phase PWM inverter applied on SVG; refer to Figures 6.10 and 6.11.

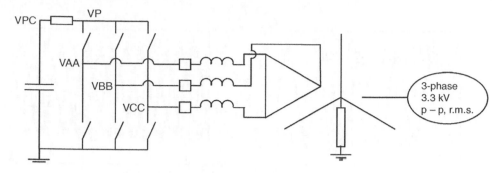

Figure 6.10 SVG circuit.

connected to a power system via inductors. Then, like a synchronous (rotating) machine compensator, the facility can be both a capacitive and inductive reactive power source. The DC source in the inverter can be replaced by a capacitor bank, since no active power is transmitted. The neutral of any side of inverter or system must be floating by this inverter circuit as shown above. So, in this case, the system side has delta-connected transformer windings, that is, floating neutral as shown in Figure 6.10. Another side of the transformer, directly connected to the outside power system, can be in any form according to the outside system's one.

Some calculated results are shown in Figure 6.11.[12] Depending on the charged voltage in the capacitor and target control voltage in the controller (i.e., TACS in this case), any capacitive or inductive reactive power is generated. For a higher inverter side voltage than the system side, capacitive reactive power is supplied to the SVG (capacitor bank mode as shown in Figure 6.11). In (a), the leading current value is approximately 500 A (crest), or approximately 2 MVA of capacitor mode operation.

As for details of the circuit parameters, see the data file.[12]

Note:

- Miscellaneous elements such as stray capacitances, snubbers, and so on, are excluded in the case. For practical cases, such are to be introduced.
- In actual cases, especially when a higher value of capacitance to earth is involved, the neutral floating system shown here may not be appropriate.
- The initialization in the calculation is complicated. In the attached calculation of cases, the initializations are not perfectly optimized.
- A relatively high capacitance value of the capacitor is necessary for the DC side. Also, a relatively high carrier wave frequency is necessary. A trial and error method seems to be suitable for surveying the matter. EMTP is one of the best appropriate tools for such a survey.

The trapped voltage in the capacitor is controlled by the phase angles difference between Vcont (inside reference voltage) and the system voltage; that is, transmitting active power via an inductor is controlled by the phase angle difference of both side voltages, resulting in the change of voltage of the capacitor.

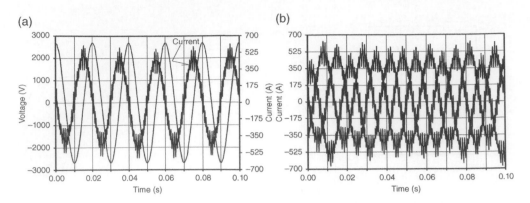

Figure 6.11 SVGs performance.[12] (a) Power system voltage and current. (b) Three-phase currents.

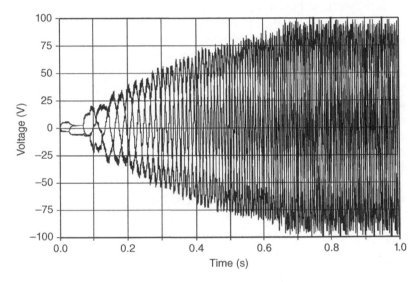

Figure 6.12 Three-phase load terminal to neutral voltages. Under linearly rising frequency/ampere reference voltage condition.[13]

By the way, in the PWM inverter shown in Figure 6.7, applying any kind of reference voltage wave shape, the relevant corresponding wave shape of voltage/current is produced in the load circuit. One example is shown in Figure 6.12 where the reference voltage wave shape is a linearly rising frequency/amplitude (variable voltage, variable frequency) one.[13] The condition may be useful for motor starting. The same circuit diagram as applied in PWM inverter is applicable to "current regulated inverter" where only the controlling principle is different.[14]

---

[13] ATPData6-24.dat: Same as data file 12, but variable voltage, variable frequency starting wave creating.
[14] ATPData6-41.dat/Data6-41.acp: Current regulated inverter circuit.

Appendix 6.A: Example of ATPDraw Picture

ATPDraw circuit diagram picture by Data6-01.acp[1] is shown in Figure A6.1. For details, also see Figure 6.1 and Data6-01.dat.[1]

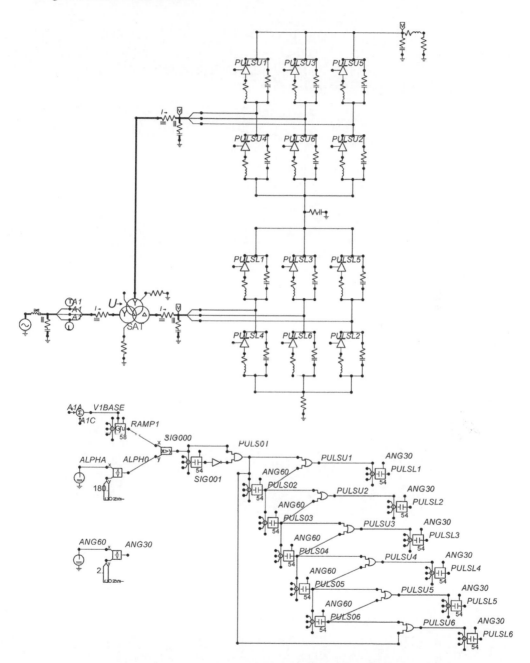

Figure A6.1   HVDC converter circuit diagram by ATPDraw.[1]

Reference

[1] E. Masada, and K. Kusumoto (1999) *Power Electronics*. Ohmsha (in Japanese).

Part II

Advanced Course-Special Phenomena and Various Applications

7

Special Switching

7.1 Transformer-Limited Short-Circuit Current Breaking

The transformer-limiting fault (TLF) is defined as an event where all the short-circuit current is delivered to a short-circuit fault point via the transformer and is interrupted by a breaker. The increased rate and peak value of the transient recovery voltage (TRV) would be greater than when interrupting the rated current. According to international standards, the amplitude factor of TRV is 1.7, which is one of the severe interrupting conditions for breakers.

Figure 7.1 shows an equivalent circuit to single-phase TLF current interrupting conditions. In Figure 7.1, the transformer is shown as a commonly used T-type equivalent circuit. When considering TRV in the current injection method, current, whose polarity is opposite to the interrupting current, will be injected from both terminals of the breaker to a circuit whose power source has been short-circuited. At that moment, the magnetizing inductance of the transformer will become parallel with the leakage impedance on the primary side of the transformer and the source impedance. In general, the magnetizing inductance of the transformer at the commercial frequencies can be ignored because it is much larger than the aforementioned impedances.

The TLF current interrupting conditions are characterized by the phenomena of commercial frequencies when the current is flowing, and after the current has been interrupted, by the TRV frequency in the order of several kilohertz to several hundred kilohertz [1].

Figure 7.2 shows the example calculation circuit for the TLF.[1,2] Calculated results are shown in Figure 7.3. The short-circuit current is 5.4 kA at its peak. TRV has a sinusoidal wave shape with a single frequency.

---

[1] DataII-7-1-1.acp
[2] DataII-7-1-2.acp

Power System Transient Analysis: Theory and Practice using Simulation Programs (ATP-EMTP), First Edition.
Eiichi Haginomori, Tadashi Koshiduka, Junichi Arai, and Hisatochi Ikeda.
© 2016 John Wiley & Sons, Ltd. Published 2016 by John Wiley & Sons, Ltd.
Companion website: www.wiley.com/go/haginomori_Ikeda/power

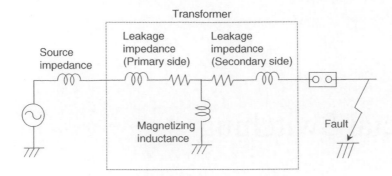

Figure 7.1   TLF equivalent circuit in a power system.

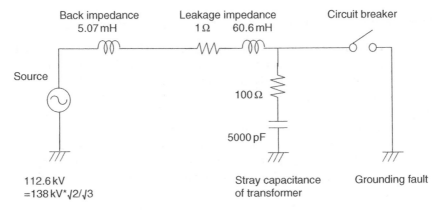

Figure 7.2   Calculation circuit for TLF.[1,2]

7.2   Transformer Winding Response to Very Fast Transient Voltage

A single-phase transformer with two windings has an iron core, low-voltage side winding, and high-voltage side winding, shown in Figure 7.4. The transformer has the self-inductance and mutual inductance in windings. Also, the transformer has stray capacitances between the iron core and the winding.

In the transformer model, that is, "TRANSFORMER," "XFORMER," or "BCTRAN" in ATP-EMTP, each winding is modeled as one inductance, where current value and voltage distribution rate are uniform along the turns. But in the high-frequency region, the voltage distribution in the winding is not uniform. It is known by the steep front of overvoltage application—the voltage stress around the entrance terminal winding is severe.

It is not easy to model the transformer in a high-frequency region because it is not easy to estimate the values of the inductances or capacitances. But the user can calculate the voltage distribution in the windings by using the high-frequency transformer model.

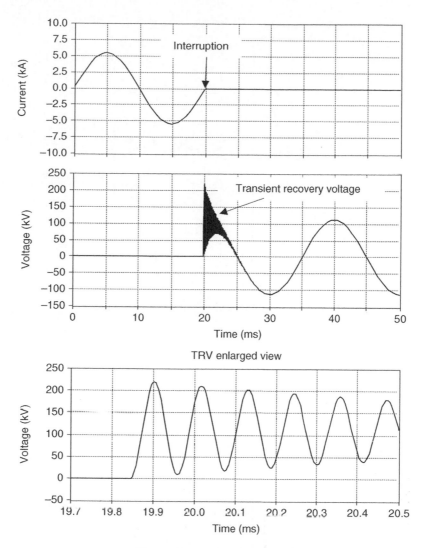

Figure 7.3 Calculation result.

Here, an air-core reactor located in a metal cylinder (earthed), detailed dimensions of which are shown in Figure 7.5, is taken up.[3,4] The reactor is divided into six sections, each consisting of five turns, having self- and each-other mutual inductances. There are also capacitances within turns and to the cylinder.

Figure 7.6 shows the calculation circuit. The inductances and resistance are modeled by mutually coupled RL (51, 52, …). The user can simulate 40 phases using the

---

[3] DataII-7-2-1.acp
[4] DataII-7-2-2.acp

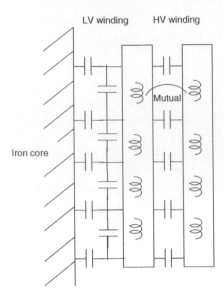

Figure 7.4   Equivalent circuit of transformer inside.

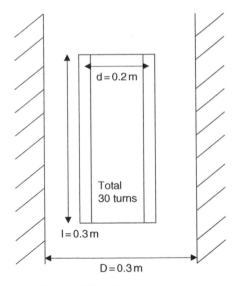

Figure 7.5   Calculation model.

mutually coupled RM model in the original EMTP. In ATPDraw, a six-phase model is prepared. When more than seven phases are used, the user must make the calculation data manually [2].

All values are shown in the data file.[3,4]

For stable transient calculation, these inductances (both self and mutual) are to be calculated as appropriately as possible. Inserting resistors of appropriate values in series to the self-inductances, which are easily introduced in mutually coupled RL elements, may bring better results.

When steep ramp and step voltage are being applied from one terminal while the other side is earthed, voltages at every five turns are shown in Figure 7.7, the enlargement of the very initial part of which is in Figure 7.8. From these, voltage stresses on the inside part are apparently delayed; that is, only the entrance part is stressed by very steep voltage stress.

In Figures 7.9 and 7.10, voltages at every five turns are shown while 1 A (crest value) of AC current in a wide range of frequency is applied. Figure 7.9 shows the voltage magnitude that corresponds to impedance in ohms.

Figure 7.10 shows phase angle corresponding to impedance phase angle. Positive corresponds to inductive. From these two figures, up to the fundamental inherent frequency(i.e., about 2 MHz)

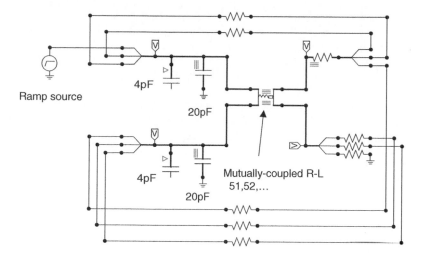

Figure 7.6 Calculation circuit.[3,4]

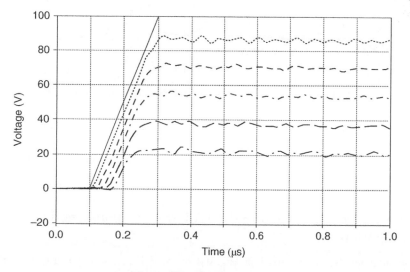

Figure 7.7 Calculated result.

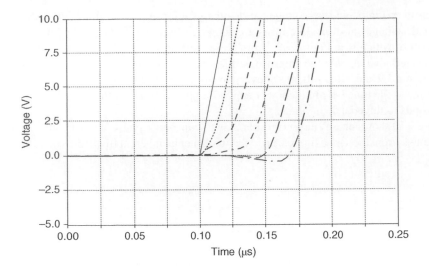

Figure 7.8   Enlarged view of Figure 7.7.

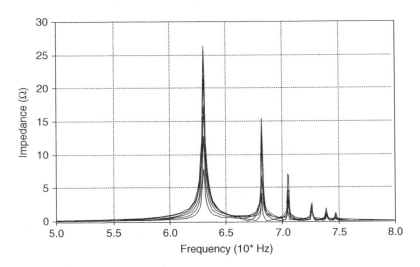

Figure 7.9   Frequency response of the coil impedance.

voltages are in a linear relation along the turns. Also, the phase angles are uniform. This means that up to that frequency, the reactor can be as one inductance.

For higher frequencies, such relations have never been kept; therefore, multi-inductance modeling is inevitable.

7.3   Transformer Magnetizing Current under Geomagnetic Storm Conditions

A severe geomagnetic storm occurred on August 4 and 5, 1972, caused by a large solar flare. The geomagnetic storm produced a large number of disturbances to electrical power systems [3–5].

DC current, due to the terrestrial magnetism change, flows in a very long transmission line in a north-south direction and reaches up to about 100 A. The transmission line terminates with a transformer at each substation.

Therefore, DC current flows through the transformer in such a circumstance up to around 100 A. The DC current mostly flows on only one side-winding of the transformer; thus, the iron core is saturated. The flux goes out of the iron core and may heat the transformer iron case due to the higher iron loss rate of the material.

In 1989, a large blackout in US and Canadian electric utilities was reported due to such a cause. Transformer magnetizing performance is shown in Figure 7.11 under such a super-imposed DC current.

By applying a certain AC voltage, the flux linkage can be any of flux (1) or flux (2), depending on the initial condition. In mathematics, the matter corresponds to the integration constant, that is, the flux is integration of the applied voltage to the inductance. The corresponding magnetizing current is current (1) or current (2). Thus any kind of current can exist. Actually, the flux bias, that is, the initial condition of the time period concerned, is fixed as a steady-state condition. As for the inrush current shown in the previous section, the initial current is of the most interest in most cases. But the phenomenon in this section lasting several tens of minutes

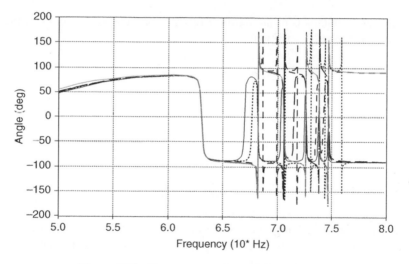

Figure 7.10 Frequency response of the coil angle.

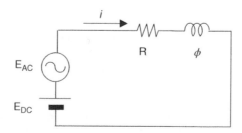

Figure 7.11 Equivalent circuit.

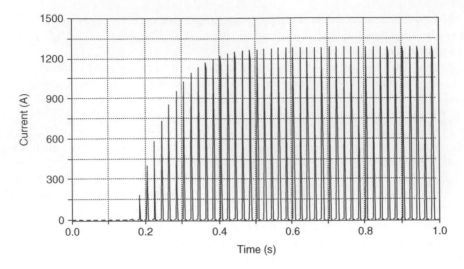

Figure 7.12 Calculated result.

at the steady-state condition is of the most interest. As for the circuit diagram in Figure 7.11, the next equations are easily obtained:

$$E_{AC} + E_{DC} = Ri + \frac{d\phi}{dt}$$

$$E_{DC}\left(t_2 - t_1\right) = \int_{t1}^{t2} Ridt$$

$$\frac{E_{DC}}{R} = \frac{1}{t_2 - t_1} \int_{t1}^{t2} idt.$$

The second equation is the integration of the first one. The third one is just a modification of the second one, which shows the average current value is just the DC current value applied.

The ATP-EMTP calculation result is shown in Figure 7.12; also see the data file for circuit parameter details.[5] In the calculation to attain a shorter time interval to the steady state, the series resistor values are intentionally enhanced. Otherwise, the time to steady state is to be several tens of seconds for the actual circuit parameters. The calculation was done applying both AC and DC voltages to the transformer without initial residual flux.

---

[5] DataII-7-3-1.acp

Note:

- A geomagnetic storm condition lasts several tens of minutes, while the thermal time constant of a transformer is on the order of one to several hours. Higher current lasting less than 1 min, such as inrush current, is of no importance as to thermal phenomena. The electrical time constant around a transformer is far less than 1 min. Therefore, electrically steady-state phenomena are of importance regarding a geomagnetic storm.
- In the data file calculating the phenomena, introducing initially residual flux and/or another timing of source voltage, a different current wave shape is obtained only for the initial time interval. After some time, the current reaches of the same steady-state value.

7.4 Four-Armed Shunt Reactor for Suppressing Secondary Arc in Single-Pole Rapid Reclosing

As the first step of studying switching phenomena in systems with four-armed shunt reactors, mathematical study seems to be beneficial to grasp the outline (see Refs. [6–8]). In single-pole rapid reclosing, where only the faulted phase of a transmission line is opened, the faulting arc is to quench during the reclosing time interval. By electrostatic coupling with the sound phases, a certain level of arc current tends to continue without quenching. As the system voltage gets higher and the transmission line gets longer, the tendency increases.

For eliminating the arc current (secondary arc current) aiming for successful reclosing, a four-armed shunt reactor where the neutral is earthed by means of another reactor is applicable. Figure 7.13 shows the circuit layout.[6]

Part (a) shows the system layout during one phase line to ground faulting, where both ends of the phase are open. A secondary arc may exist.

Part (b) shows the equivalent circuit at the faulting point, where, assuming voltages along the phase v and w lines are quasi-uniform, voltages are applied from the point instead of both ends, that is, $e_u = 0$ (faulting), e_v, and e_w. i_u is the secondary arc current. Z_0, Z_1, and Z_2 are sequence component reactance of the line section (capacitances and inductances of four-armed shunt reactor shown in (c) connected in parallel). The following equations are obtained:

$$\begin{bmatrix} e_0 \\ e_1 \\ e_2 \end{bmatrix} = \frac{1}{3} \begin{bmatrix} 1 & 1 & 1 \\ 1 & a & a^2 \\ 1 & a^2 & a \end{bmatrix} \begin{bmatrix} 0 \\ e_v \\ e_w \end{bmatrix}$$

and

$$\begin{bmatrix} i_0 \\ i_1 \\ i_2 \end{bmatrix} = \begin{bmatrix} e_0 / Z_0 \\ e_1 / Z_1 \\ e_2 / Z_2 \end{bmatrix},$$

---

[6] DataII-7-4-1.acp

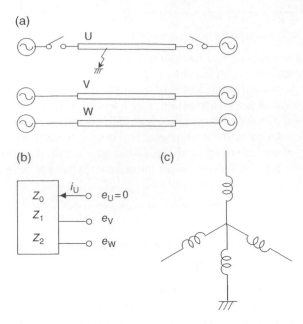

Figure 7.13 Four-armed shunt reactor for suppressing a secondary arc in single-pole rapid reclosing.

where $a = -\dfrac{1}{2} + j\dfrac{\sqrt{3}}{2}$

Except for rotating machines, in transmission systems, $Z_1 = Z_2$. In a transmission line with a four-armed shunt reactor, parameters other than the neutral reactor's are fixed by the relevant system condition. So adjusting the neutral reactor reactance value, we can have

$$Z_0 = Z_1 = Z_2.$$

Introducing this condition, then we can have $i_u = 0$, applying these equations; that is, the secondary arc current can be suppressed (Figures 7.14 and 7.15).

Note:

During the switching of such a transmission line, due to the nonlinearity of the reactors (as usually iron cores are used), certain values of transient voltages appear at the neutral point and some insulation failures have been reported. For sophisticated insulation design, especially around the neutral point, accurate analysis introducing every detailed parameter of the system, including the nonlinear characteristics of iron cores, is recommended.

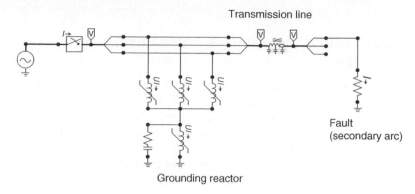

Transmission line

Fault
(secondary arc)

Grounding reactor

Figure 7.14 Calculation circuit.[6]

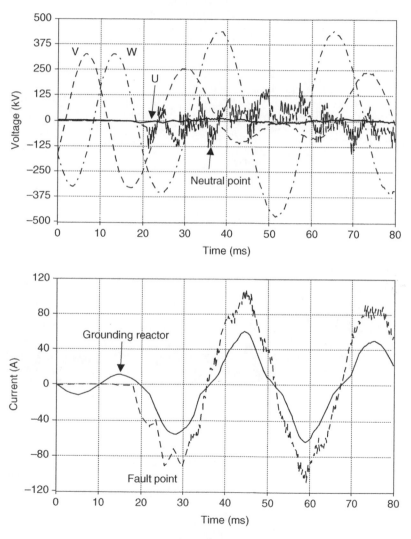

Figure 7.15 Calculated result.

7.5   Switching Four-Armed Shunt Reactor Compensated Transmission Line

When switching a transmission line with four-armed shunt reactor compensation, the purpose of which is to suppress secondary arc current when single-phase reclosing, due to unbalanced saturations of the shunt reactor arms, overvoltages appear at the neutral point, the voltage of which point is zero in the steady-state condition (see Refs. [6–8]). The following is the most simplified example of the phenomenon.

As shown in Figure 7.16, 400 kV 300 km of overhead transmission line with general parameters is compensated by a four-armed shunt reactor, the compensation ratio of which is 60%.[7] The no-load line is energized from the left end and then dropped.

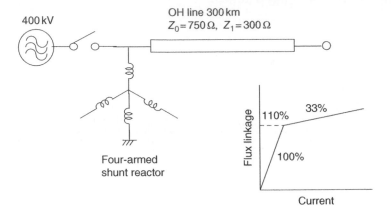

Figure 7.16   Calculation circuit.[7]

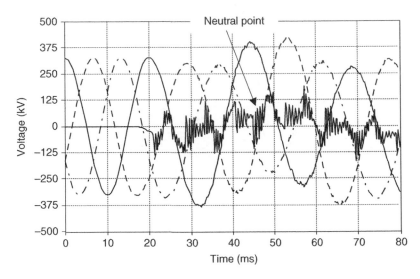

Figure 7.17   Calculated result.

[7] DataII-7-5-1.acp

Such a reactor is generally of a gapped-core type, so the saturation characteristic is assumed as shown in the figure. Some nonlinear elements dominate the phenomena, so digital calculation seems to be the most applicable.

As for the details of the modeled parameters, see the data file.[7]

ATP-EMTP calculation results for the shunt-reactor terminal and neutral voltages when the line is dropped are shown in Figure 7.17. In this case, a significantly high voltage appears at the neutral point of the reactor after the line dropping, which may be very important for reactor insulation design.

Care should be taken, as the phenomenon much depends on the relevant system parameters— details of the transmission line parameters, shunt reactor compensation rate, shunt reactor saturation characteristics, and so on. Modeling that is as precise as possible is necessary for the actual case evaluation.

References

[1] S.R. Lambert (1993) Circuit breaker transient recovery voltage. IEEE PES, IEEE Tutorial Course, Application of Power Circuit Breakers, 93EH0 388-9-PWR, pp. 32–37.
[2] *ATP-EMTP Rule Book*. Canadian-American EMTP Users Group.
[3] V.D. Albertson, J.M. Thorson (1974) Power system disturbances during a K-8 geomagnetic storm: August 4, 1972, *IEEE Transactions on Power Apparatus and Systems*, **PAS-93**, 1025–1030.
[4] V.D. Albertson, J.M. Thorson, S.A. Miske (1974) The effects of geomagnetic storms on electric power systems, *IEEE Transactions on Power Apparatus and Systems*, **PAS-93**, 1031–1044.
[5] W.A. Radasky, J.G. Kappenman (2010) Impacts of geomagnetic storms on EHV and UHV power grids. *2010 Asia-Pacific International Symposium on Electromagnetic Compatibility, Beijing, China*, April 12–16, 2010.
[6] B.R. Shperling, A. Fakheri, and B.J. Ware (1978) Compensation scheme for single-pole switching on untransposed transmission lines. *IEEE Transactions on Power Apparatus and Systems*, **PAS-97** (4), 1421–1429.
[7] B.R. Shperling, A.J. Fakheri, C.H. Shih, B.J. Ware (1981) Analysis of single phase switching field tests on the AEP 765 kV system, *IEEE Transactions on Power Apparatus and Systems*, **PAS-100** (4), 1729–1735.
[8] H.N. Scherer, N.N. Belyakov, B.R. Shperling, V.S. Rashkes, J.W. Chadwick, K.V. Khoetsian (1985) Single phase switching tests on 765 kV and 750 kV transmission lines, *IEEE Transactions on Power Apparatus and Systems*, **PAS-104** (6), 1536–1548.

8

Synchronous Machine Dynamics

In the middle of the 1980s, the present Type 59 synchronous machine model program was implemented into the Electromagnetic Transients Program (EMTP) and put into practical use. Then, in the mid-1990s, the Type 58 model, which was significantly improved over the former, was also put into practical use in ATP-EMTP [1]. Most power sources in AC power systems are synchronous generators, so the dynamics of the machines are, sometimes, of great interest, especially regarding relatively short time intervals of phenomena. Only time domain analysis is applicable to such fast transient phenomena, occurring in only milliseconds. In such circumstances EMTP is a significantly useful tool for analyzing power system dynamics. As a special feature of Type 58, calculations are stable, especially in asymmetrical circuit conditions such as nontransposed overhead transmission lines, which are mostly applied in today's power systems. Furthermore, it should be noted that the present Type 59 model produces erroneous results in cases introducing field coil saturation characteristics.

The usage is mostly common to both types, except for the writing of "58" or "59" at the machine part of the program data file. In this chapter, therefore, explanations are mostly for Type 58.

8.1 Synchronous Machine Modeling and Machine Parameters

What is written in Chapter 7 of the *Rule Book, Dynamic Synchronous Machine*, is not perfectly updated, so in this chapter, updated data file coding is shown.

Synchronous machine calculation in ATP-EMTP is based on the modeling shown in Figure 8.1 (2P machine). Two coils in each of the d- and q-axes represent the rotor. While in the stator, in Type 59, three-phase coils are transformed into two coils in d- and q-axes (Park-domain, DW, QW), whereas in Type 58, three-phase coils are applied as they are. The basic equations based on the figure for Type 58 modeling are as follows.

Power System Transient Analysis: Theory and Practice using Simulation Programs (ATP-EMTP), First Edition.
Eiichi Haginomori, Tadashi Koshiduka, Junichi Arai, and Hisatochi Ikeda.
© 2016 John Wiley & Sons, Ltd. Published 2016 by John Wiley & Sons, Ltd.
Companion website: www.wiley.com/go/haginomori_Ikeda/power

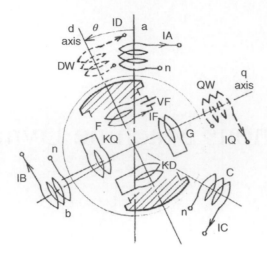

Figure 8.1   Synchronous machine modeling. DW and QW are applied only in Type 59.

As for the voltage of each of seven coils,

$$V_j = -R_j i_j - \frac{d}{dt} L_{jk} i_k,$$ (8.1)

where j, k = a, b, c, F, G, KD, and KQ, and L_{jk} = time varying functions, depending on the angle between rotor and stator.

As for the torque,

$$T = \sum_j i_j \sum_k i_k \frac{d}{d\theta} L_{jk}.$$ (8.2)

These differential equations are numerically calculated.

In practical cases, as for these machine constants, they can seldom be known by measuring or calculation. Therefore, in most cases, general machine data such as transient/subtransient reactance values/time-constants are input, and these constants are automatically calculated in EMTP. Some assumptions and simplifications are introduced during the calculation; still, the calculation is considered sufficiently accurate and reliable.

Note:

- For further details as to the coding, see *Dynamic Synchronous Machine*, Chapter 7, and the data files. *EMTP Theory Book* is also very useful for understanding.
- In using ATPDraw, the same items shown in Table 8.1 are, in principle, to be input.
- In the second and third lines from the first "58" writing, only "58" and node names are to be written. Voltages, frequencies, and angles are automatically introduced for symmetrical three-phase AC.

- A PARAMETER FITTING value of ≤2.0 means open-circuit time constants are to be used.
- A PARAMETER FITTING value of ≥2.1 of means short-circuit time constants are to be used. Generally, short-circuit time constant usage is more appropriate, excluding the saturation effect.
- A "1" in column 7, line 5, means metric unit mechanical constants are to be used. With "0" or blank, imperial units are applied.
- For resistors and reactance values, per-unit values (machine rating bases) are to be applied.
- For time constants, "second" is to be used as the unit.
- If XCAN (Caney reactance) can be applied, transient rotor coil currents such as during short-circuiting are more correctly calculated. For armature currents, little influence is introduced. Without introducing the value, XL value is automatically introduced as XCAN.
- After the blank card, writing 11, 21, 31, and 51 in Output ordering cards yields full output for one mass machine case and is generally recommended.
- For the initial condition fixing in Type 58 machine, the FIX SOURCES command is applicable in a general three-phase symmetrical case. NEW LOAD FLOW (old CAO LOAD FLOW) is also applicable, but is thought to introduce better results, especially in very limited asymmetrical circuit conditions. For details, see the BENCHMARK DCN20.DAT file.

Typical data coding format for Type 58 is shown in Table 8.1.[1]

8.2   Some Basic Examples

Figure 8.2 shows the Inf 2-Gen tandem power system. Applying the layout, some cases of calculations are made. In the system layout, two generators are connected to the transmission line via step-up transformers. The transmission line is a nontransposed double circuited one, where the phase location of each circuit is not the same for obtaining better symmetry as a whole, for example, phase a, b, c from the top in one circuit and c, a, b in another circuit in this case. Infinitive capacity of source can be represented by voltage source. Step-up transformer winding connections are delta in generator (low-voltage) sides and Y in transmission line (high voltage: HV) sides. For details, see the previous chapter on transformers.

8.2.1   No-Load Transmission Line Charging

In the first example, the no-load overhead line charging current is calculated. In Figure 8.2, by disconnecting the No. 2 generator as well as the infinitive capacity of voltage source, only the No. 1 generator supplies capacitive reactive charging current. It should be noted that capacitive charging current phenomena are in the power frequency. The overhead line parameters are, more or less, frequency dependent, so parameters are to be calculated on a power frequency basis. For details, see the relevant data file for transmission line constant calculation (LINE CONSTANTS).[2]

---

[1] Data8-00.dat: 1300MVA, 4P, 19 kV generator and step-up transformer parameters.
[2] Data8-01.dat: 300 kV, 410 mm**2 X2, 50 km Line Constants parameter calculation data.

Table 8.1 No. 58 model data coding.

```
C ==================================================================
C    3-PHASE SYNCHRONOUS GENERATOR     1300MVA-19kV-50Hz, 4P
C ==================================================================
C BUS |V|AMPLITUDE|FREQ.HZ  |ANGLE.DEG|    |START.SEC|STOP.SEC
C *-----||||------------------||-------------||----------------||----||--------||----||--------
58VG1A  15513.        50.0              −30.
58VG1B
58VG1C
PARAMETER FITTING   2.1     {Short-circuited Time Constants applied.
C |G E||NP| SMOUTP || SMOUTQ ||  RMVA  ||  RKV  ||AGLINE||  S1  ||  S2
  1 1 1   4   1.0       1.0       1300.     19.0    −1600.   1980.   3600.
                        1.0       1.2       1.0      1.2     −1.  {Same as d−axis
C   RA  ||    XL    ||   XD  ||   XQ  ||  XD'  ||  XQ'  ||   XD'' ||  XQ''
  0.0025     0.18      1.51      1.75     0.42    .55        .30       .28
C  TDs' ||   TQs'  ||  TDs'' || TQs'' ||  X0  || RN    ||  XN  || XCAN
   2.52     0.18       .042    .0313     .17     .0191     5.
C  MASS  |(EXTRS) ||  (HICO) || (DSR)  || (DSM) ||(HSP)   ||(DSD)  |
  1          1.0        1.5       0.0       0.0      0.0       500.
BLANK CARD ENDING MACHINE DATA
C OUTPUT REQ. Full Ordering ---------------------------------------------
   11
   21
   31
   51
BLANK
C CONNECTION TO TACS
C 71EX1                              {For TACS Controlling Field Exciter
   FINISH
BLANK CARD ENDING SOURCE CARDS
C |----------||------||------ ||------------||-----------|== NEW LOAD FLOW REQUEST ==
C 1VG1A  VG1B  VG1C       1.17E06  10970.
C BLANK CARD ENDING FIX-SOURCE / NEW LOAD FLOW CARDS
C |----------||------||------ ||------------||-----------||------||------||------||------||------||------||------||------|
  VG1A  VG1B  VG1C  VHT1A VHT1B VHT1C
BLANK CARD ENDING (VOLTAGE) OUTPUT
BLANK CARD GENERAL
BEGIN NEW DATA CASE
BLANK
```

In the present state, one generator supplies capacitive current to no-load transmission line(s). Fixing the generator's terminal three-phase voltages (amplitudes, frequencies, and initial phase angles), then all parts of the system voltages/currents are to be fixed. Therefore, in the present EMTP DAT file for the calculation, only the generator terminal voltage conditions are written for the initialization purpose. Also see the relevant data file for the details.[3]

---

[3] Data8-02.dat: Generator supplying transmission line charging current, 300 kV, 100 km, 2 cct.

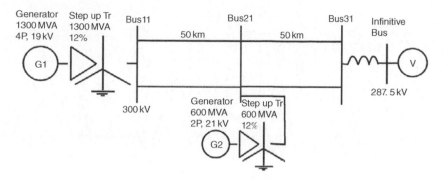

Figure 8.2   INF 2-GEN power system.

For obtaining symmetry as much in the untransposed transmission lines, the phase location in the conductors is not in a uniform order (also see the relevant data file). The time interval in this case is 0.1 ms. Generally, the generator model and transmission line length restrict the maximum allowable time step value in EMTP (0.05–0.1 ms is generally appropriate).

Some calculated results are shown in Figure 8.3.

(a) The generator's terminal voltages are symmetrical as specified in the relevant DAT. Nevertheless, the supplying currents are, more or less, asymmetrical, perhaps due to the asymmetry of the line (untransposed).
(b) Line entrance voltages are also symmetrical, but the currents are a little bit asymmetrical. The asymmetry rate is not so significant at the generator terminal, perhaps due to the transformer winding connection (Δ–Y) and the phase location of the transmission line.
(c) Currents in the rotor part are shown. As for the steady state, only the field current exists.
(d) The air-gap torque shows low amplitude of fluctuation. In perfectly symmetrical steady state condition, the torque is to be constant. Due to the asymmetry, as mentioned previously, the fluctuation is created. The average value of the torque is in the order of 0.02% of the rated one corresponding to the losses in the circuit. Mechanical losses do not concern the air gap torque in this case.

8.2.2   Power Flow Calculation

Assuming that, in Figure 8.2, G2 is disconnected and only G1 is supplying full load current to the right-hand side end infinitive source, the coding of DAT file is to be written as the relevant data file.[4] For fixing the initial power flow condition in the case, the FIX SOURCES command, written at the upper part of the file, is applied where the active power value and the generator terminal voltage are specified in the data file. Other methods, for example, specifying active and

---

[4] Data8-03.dat: One generator plus infinitive bus case, load flow calculation, FIX SOURCES option applied.

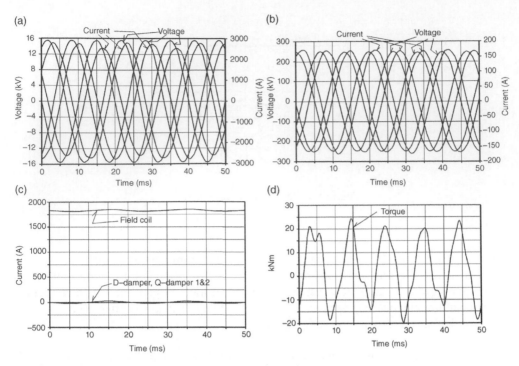

Figure 8.3 Variables in no-load transmission line charging condition. (a) Generator terminal V and C. (b) Transmission line V and C. (c) Rotor currents. (d) Air gap torque.

reactive powers, voltage, and phase angle, are also applicable. For the details, see the *Rule Book* (chapter on Load Flow).

In this case, the generator is supplying 1170 MW of active power under the rated terminal voltage condition. Some calculated results are shown in Figure 8.4.

(a) From the generator terminal voltage and current, the apparent power is calculated as approximately 1200 MVA. The current is slightly lagging.
(b) The current into the transmission line, on the other hand, lags less, due to the existence of inductive reactance between the generator and transmission line, that is, the transformer.
(c) Voltages at the locations from the generator to the transmission line end show step by step lagging in the voltage phase angles along the line. Please note, the base of the voltage phase angle at the generator is delayed by 30° compared with the transmission line's one due to the transformer winding connection (Δ–Y connection).
(d) From the air gap torque, the active power output is calculated to be 1160 MW, which is close to the specified value (1170 MW); thus, the calculation seems to be accurate.

In the next example, two generators supply load currents to the remote infinitive source. In Figure 8.2, all components are connected. For initializing the system, the FIX SOURCES

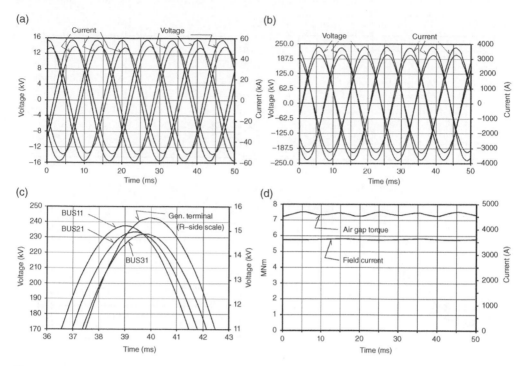

Figure 8.4 Variables under generator full load supplying condition. (a) G1 terminal voltage and current. (b) Transmission line voltage and current. (c) Generator terminal and line voltages. (d) Air gap torque and field current.

menu is also applied. The active power values and terminal voltages for the two generators are specified; for details, see the relevant data file.[5]

Some calculated results are shown in Figure 8.5.

(a) The generator terminal voltages of G1 and G2 shows phase angle difference according to the load flow current in between. The apparent output powers are calculated to be approximately 1280 and 650 MVA for G1 and G2, respectively.

(b) The voltage and current phase angles at BUS/Transmission parts are advanced by approximately 30° compared to ones of the generator terminals, due to the relevant step-up transformer winding connections.

(c) Comparing BUS11–BUS21 and BUS21–BUS31, the voltage phase angle difference is bigger in the latter. The reason is due to the higher active load current in the relevant transmission line reactance, where the reactance of the two sections is equal.

(d) From the torques of G1 and G2, the active powers are calculated to be 1160 and 620 MW, respectively. Also, the rotor d-axis physical angles are shown. It should be noted that G1 is a 4P machine (1500 r.p.m.) and G2 is a 2P one (3000 r.p.m.) when calculating the active power or observing the rotor angle position.

---

[5] Data8-04.dat: Two generators plus infinitive bus case, full load flow calculation by FIX SOURCES.

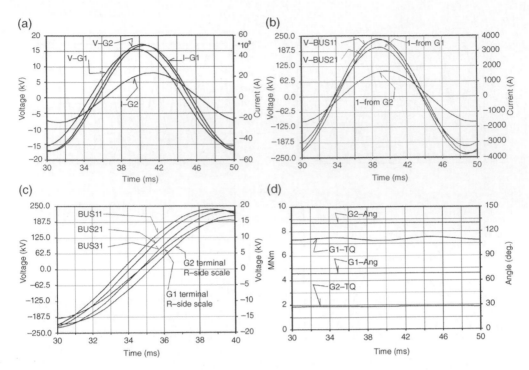

Figure 8.5 Variables while two generators are feeding load currents. (a) Generator terminal voltage and current. (b) Transmission line voltage and current. (c) Line and generator terminal voltages. (d) Generator TQ and d-axis angle.

8.2.3   Sudden Short-Circuiting

Peculiar phenomena appear when short-circuiting the circuit close to a synchronous generator. Only a time-domain calculation program, such as EMTP, is applicable for such analysis.

Under the condition of the two generators supplying full load current in Figure 8.2, a three-phase simultaneous earthing fault in one line circuit near BUS11, which will be cleared after approximately three cycles, is considered.[6] The calculated results are as follows (see Figure 8.6):

(a) The total of the three-phase faulting currents is shown. No delayed current zero crossing occurs.

(b) Generator terminal output currents are shown. No delayed current zero crossing occurs here either.

(c) In the faulting location voltages, it should be noted that at the fault clearing, the first phase to clear voltage is much damped due to the X0/X1 < 1.0 condition regarding the short circuit reactance at the location.

(d) During the faulting transients, tremendously high currents flow in the rotor. Please note that each current is converted to the same turn number of coils as the field one.

---

[6] Data8-05.dat: Three-phase earthing fault near No. 1 plant step-up transformer HV side under the same system conditions, three-phase simultaneous faulting.

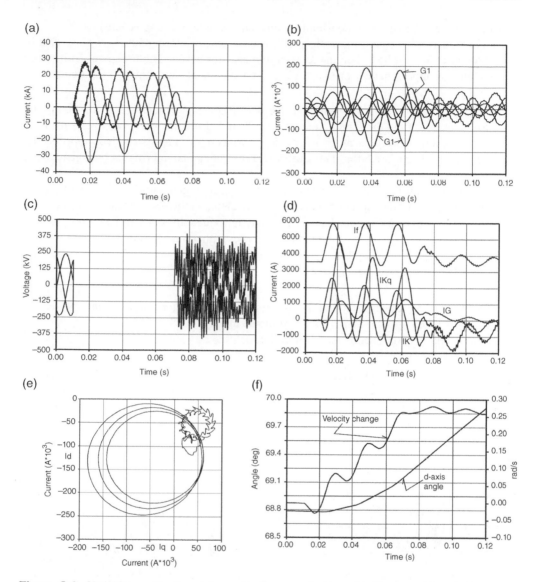

Figure 8.6 Variables under three-phase simultaneous faulting. (a) Total earthing fault current. (b) Generator load and fault current. (c) Fault location voltage. (d) Generator rotor currents. (e) Armature current in the d-q domain. (f) Rotor d-axis angle and velocity change.

(e) In the d-q domain diagram, the circle radius and the distance of the center position from the zero point correspond to armature current's DC and AC components, respectively. Then each damping rate is clarified visually.

(f) During short-circuiting, due to less active power output, the rotor accelerates and the rotor angle advances.

In the next case, three-phase nonsimultaneous fault initiation timing is taken up.[7]
The fault condition is the same as the previous case, except for the faulting timing. That is, first two phases are earthed, then the remaining phase is earthed. The time interval between the two faults is approximately one half cycle. The main results are shown in Figure 8.7.

(a) The fault position is the transmission line near BUS11 in Figure 8.2. In the total fault current, current from the remote source is involved. During the initial two cycles from the fault initiation, delayed current zero crossing appears in one phase, which disappears after three cycles. Therefore, after three cycles, the current can be interrupted by a general AC circuit breaker.

(b) The figure shows the current from only the adjacent generator's side. In this case, noninterruption of the fault current is calculated. Here, a more significant delayed current zero crossing appears in one phase, which may introduce difficulty if the current is to be interrupted by the relevant circuit breaker. These phenomena relate faults closely to generator plant(s) and step-by-step three-phase faulting (evolving fault). From a statistical point of view, this may be a rare occasion.

In the next example, an earth fault at the location between the generator G1 and the step-up transformer is taken up. The main results are shown in Figure 8.8.
The three-phase simultaneous fault is introduced.

(a) The total fault current involving one from the HV side shows a less significantly delayed current zero crossing. After the interruption of the current from the generator, that is, by a generator circuit breaker, current from the HV side remains, which would be, in an actual case, interrupted at the HV side.

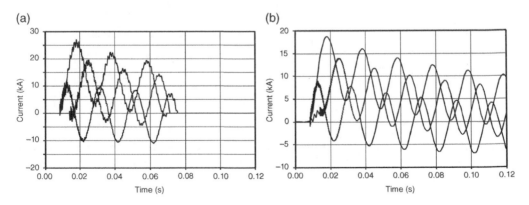

Figure 8.7   Three-phase nonsimultaneous faulting. (a) Fault current (interrupted). (b) Generator supplying current.

---

[7] Data8-06.dat: Same as previous, but with on-load condition and nonsimultaneous three-phase faulting; minimum DC in the fault current.

(a)

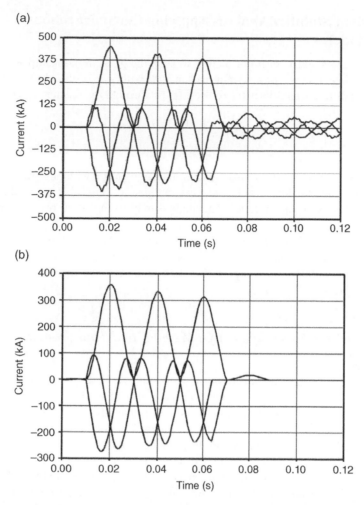

(b)

Figure 8.8 Earthing fault at the generator terminal. (a) Total fault current. (b) Current from G1.

(b) Current from only the generator side is shown, which is interrupted by a (generator) circuit breaker at the location. Due to the more significant delayed current zero crossing compared to the total current, interruption in one phase is delayed by one cycle. Depending on the generator and system circuit parameter(s), a much more significant delayed current zero crossing may occur.[8] This is a typical topic regarding generator circuit breakers.[9]

---

[8] Data8-07.dat: Similar to the 8-05 case, but with nonsimultaneous three-phase faulting to create maximum zero skipping fault current.

[9] Data8-0x.dat: Step-up transformers with magnetizing linear reactances (no-saturation) are applied. Reference only.

8.3   Transient Stability Analysis Applying the Synchronous Machine Model

In AC power systems, each generator there is to keep the phase relationship according to the relevant power flow, that is, for a certain reactance X, both side terminal voltages V_1 and V_2, and phase angle difference θ, the active power flow P through the reactance is calculated as

$$P = \frac{V_1 \cdot V_2 \sin \theta}{X}. \tag{8.3}$$

By disturbances such as short-circuits, sudden load rejection, switching transmission lines, and so on, each generator may accelerate/decelerate due to the probable unbalance between the driving and load (air gap) torques. The angle θ may swing; such phenomena are called "transient stability."

In this section the phenomena are explained mainly applying the time domain analysis, contrasted with the conventional process "power frequency phasor domain analysis" or the classic "equal-area method." Also explained is the countermeasure to enhance the power flow limit to keep stability.

8.3.1   Classic Analysis (Equal-Area Method) and Time Domain Analysis (EMTP)

First, let's check the conventional phasor domain analysis based on the classic equal-area method by cross-checking with time domain analysis, or ATP-EMTP.

Introducing the so-called "one generator versus infinitive bus system" shown in Figure 8.9, the basic phenomena are to be surveyed. Other parameters not shown in the figure are to be referred to in the data file(s) of this chapter. In applying Equation (8.3), what is V_1, the sending source voltage? This seems to be an imagined (virtual) inside voltage of the machine and may be a changeable one. As shown later, the flux linkage of the armature winding is kept relatively constant during the relevant phenomena. The voltage, corresponding to the flux, seems to be appropriate as this V_1. Therefore, as the reactance of the machine under the phenomena, the armature winding leakage seems to be properly applied. For the transmission line reactance, parameters of a fully transposed one are applied in the case for simple calculation purposes. (Double-circuited line parameters applied.)

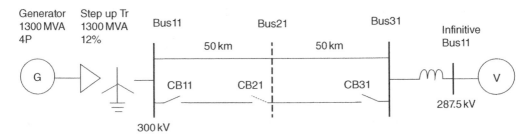

Figure 8.9   One generator versus infinitive bus system.

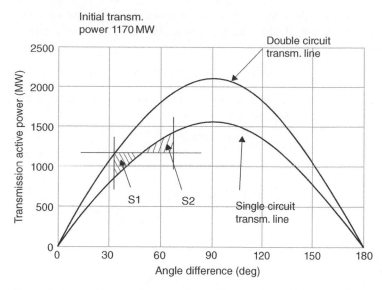

Figure 8.10   Transmission active power characteristics for a single-circuited and double-circuited line.

In Figure 8.10, transmission active power versus angle difference for single-circuited and double-circuited lines is shown. As the sending source voltage V_1, 300 kV (corresponding to the generator's rated terminal voltage) is applied, though, quite correctly speaking, the inside voltage is more or less higher than the terminal voltage. Thus, V_2 of 287.5 kV is applied.

For a double-circuited line transmitting 1170 MW (90% of power factor of 1300 MVA), the initial angle of 32.5° is calculated by the equation. Also, for single-circuited line transmitting, 48.5° is calculated. These values, of course, agree with Figure 8.10.

Assuming sudden switching over from double-circuited to single, in Figure 8.10, the vertical coordinate corresponds to the machine load torque and, therefore, the area, (i.e., the product of torque and angle shifting) corresponds to energy. Therefore, S_1 is excess energy from the turbine up to the steady-state point in the single-circuited line. Then the generator is accelerated and the angle advances up to the angle corresponding to $S_1 = S_2$. This is the simplest case of an equal-area method. The maximum angle for overswing of approximately 69° is graphically obtained for this case.

Next, let's check these values quantitatively applying EMTP (time domain analysis program). The system and machine parameters coincide with ones in the previous section, but as a small difference, untransposed line parameters are applied when considering general cases.[10]

The machine's rotor angle position is shown in Figure 8.11.[11] Care should be taken (in EMTP) that the angle corresponds to the actual pole's geometrical one for a four-pole machine based on the infinitive bus voltage electrical angle, which means 1 multiplied by 2. Also, the

---

[10] Data8-100.dat: 1G (No. 58 SM) vs. infinitive bus, full loading initialized by FIX SOURCES.
[11] Data8-101.dat/Draw8-101.acp: Same as data file 9, but with one circuit of double-circuited transmission line dropping.

machine is connected to the system via delta-star windings of the step up transformer, which means plus 30°. So, 66° for a double-circuited line steady-state case is calculated, corresponding with the normalized electrical angle basis, to be

$$(66 \times 2)(4P - 2P) + 30(Delta - Star) = 162. \tag{8.4}$$

Figure 8.12 shows the flux vector linked with the armature winding, based on a d-q-axis plane. The armature flux position angle is then

$$162 - 39 = 123°. \tag{8.5}$$

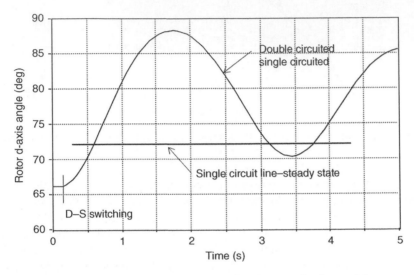

Figure 8.11   Rotor d-axis position base on infinitive bus voltage angle via star-delta step up Tr, for a four-pole machine.

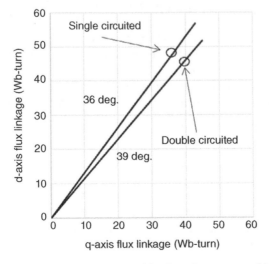

Figure 8.12   Armature winding flux linkage position based on rotor position—electrical angle.

Therefore, the voltage angle at the armature winding is (as $d\phi/dt = -V$ for a generator)

$$123 - 90 = 33°. \tag{8.6}$$

The value agrees well with the former hand-calculated one in the classic method.

For a single circuit, by a similar calculation, 49° is obtained, which is also in good agreement with the hand-calculated one also shown previously.

As for overswinging (Figures 8.11 and 8.15), the instant of maximum swing in Figure 8.11, the flux angle (not shown), is approximately 45°; therefore, the maximum voltage angle is

$$88 \times 2 + 30 - 45 - 90 = 71°, \tag{8.7}$$

which shows fairly good agreement with the equal-area method result.

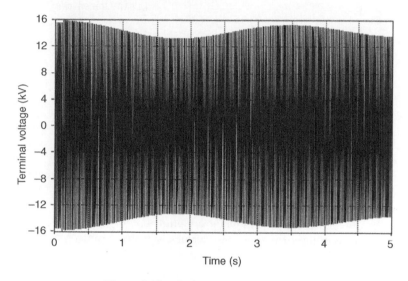

Figure 8.13   Generator terminal voltage.

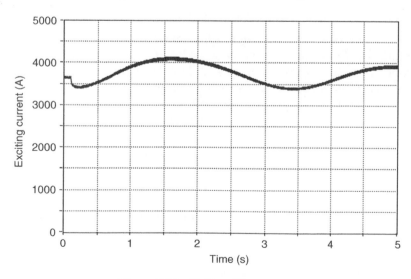

Figure 8.14   Field exciting current.

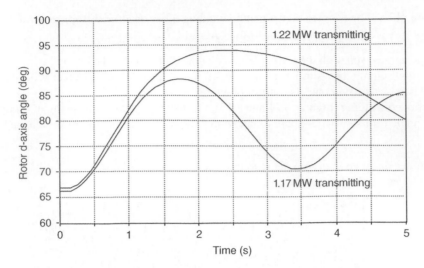

Figure 8.15   Rotor angle by overloading.

Note:

In classic calculation methods such as the equal-area method, both mechanical and electrical source points are assumed to be based on a common point. Actually, the mechanical energy transferring point is the rotor, but the electrically transferring point is vague. In this section, values on the armature winding are considered, although in general cases it is in approximation. Also, values on the armature terminal voltage are applied.

8.3.2   Detailed Transients by Time Domain Analysis: ATP-EMTP

In Figure 8.11, the maximum rotor swing is approximately 20° for a four-pole machine, that is, electrically 40°. In Figure 8.10, the value is 36°. The difference depends on the flux movement on the rotor, as shown in Figure 8.12. In a classic method, such as the equal-area method, the source-side voltage is assumed to be constant.[11]

In time domain analysis, every actual value is calculated. Also, actual voltage amplitude is changeable, which it cannot be in the classic method. Some details are shown in Figures 8.13 and 8.14, the former of which is generator terminal voltage and the latter is field exciting current. Both change during the transient time interval. In Figure 8.12 the flux angle and amplitude changes are shown. These show that the variables of the generator are variously changeable during the transient.

As the next example, increasing the transmission active power by 5%,[12] the rotor angle swing is shown in Figure 8.15 in comparison with the former case.[11,12] The amplitude increases and, furthermore, the recovery is much delayed.

---

[12] Data8-102.dat/Draw8-102.acp: Same as previous, but under 105% overloading.

The condition seems to be critical. Actually, though not shown here, the generator loses the synchronism by 5% of transmission power enhancement (out-of-phase).

In the next example, a line earthing fault followed by one circuit of the double-circuited line opening is taken up. Two cases are introduced.

Case 1: Excluding the intermediate switching station in Figure 8.16, faulting (F) is cleared by CB11 and CB31, that is, the whole length of the line becomes single circuited. The generator output (transmission active power) of 1.12 GW is applied.[13]

Case 2: introducing intermediate switching station, the fault is cleared by CB11 and CB21, that is, only half of the line is single circuited.[14]

The results are shown in Figure 8.17 as the rotor angle swing. In case 1, 1.17 GW of transmission power results in out-of-phase so, as the critical value, 1.12 GW is introduced. In case 2, where

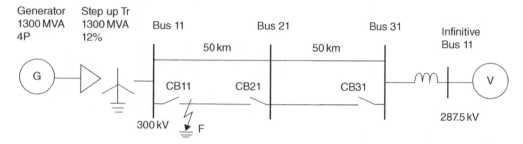

Figure 8.16   Power system diagram, 3LG fault—one circuit clearing.

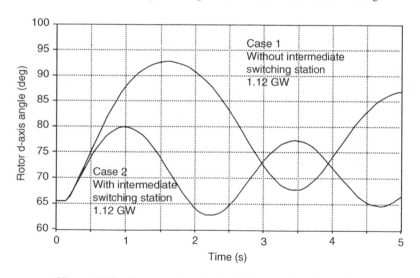

Figure 8.17   Rotor angle swing by 3LG—one circuit clearing.

---

[13] Data8-103.dat/Draw8-103.acp: Same as the previous, but three-phase earthing fault and clearing by dropping one circuit (full length), under a critical loading condition (96%).

[14] Data8-104.dat/Draw8-104.acp: Same as the previous, but half length of transmission line, under a full loading condition.

only the half of the line is single circuited, due to less enhancement of the line impedance, significant improvement is apparent as for transient stability enhancement.

In the next case, the second generator unit is connected to the intermediate bus shown in Figure 8.18, the rating of which is 600 MVA with a two-pole machine, feeding 600 MW.[15] The faulting and clearing sequence is the same as the former.

The result (rotor angle swing) is shown in Figure 8.19. For easy comparison of the angle swing between two-pole and four-pole machines on the same basis, the rotor angle of a two-pole machine is multiplied by 0.5 in the plot. As the faulting spot is remote from the second generator, the angle swing is far less than the first one. For a two-generator case, though the initial angle is higher than a one-generator case due to the higher total power flow, the stability state is

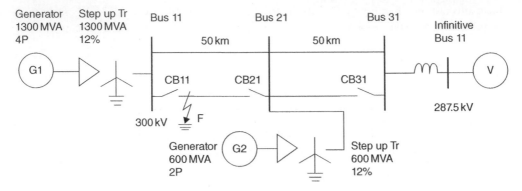

Figure 8.18   Two generators versus the infinitive bus case.

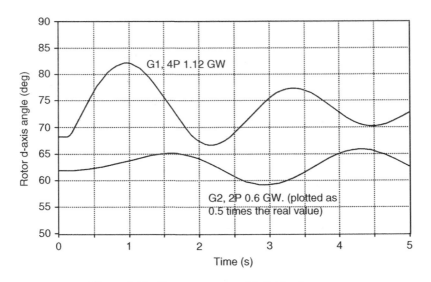

Figure 8.19   Two generator and infinitive bus case.

---

[15] Data8-105.dat/Draw8-105.acp: 2G versus the infinitive bus system, three-phase earthing fault, and clearing by dropping one circuit (half length) of the line.

almost the same. In a two-generator case, the swing of the first generator seems to transfer to the second one, that is, the second one's swing amplitude enhances. Much more complicated interaction phenomena may occur in multimachine cases.

8.3.3   Field Excitation Control

Figure 8.13 shows the terminal voltage is damped when the angle enhances. From Equation (8.3), the lower voltage corresponds to lower transmitting power; thus, the generator, due to the lower load, tends to accelerate. By increasing the excitation current, resulting in higher terminal voltage, the transmitting power enhances, and the machine acceleration is damped. Generators have an AVR (automatic voltage regulator) and/or a PSS (power system stabilizer), the main purpose of which is to keep the terminal voltage constant. But in introducing special control of them, enhancement of transient stability is also expected. PSS is especially designed for the high capacity of modern machines.

Figure 8.20 shows the block diagram of AVR/PSS applied to the generator, which is of a thyristor type with very rapid response. As the source voltage to drive the AVR/PSS is usually supplied by rectifying the generator output voltage V_t, the output is proportional to the generator terminal voltage V_t, as shown in the figure.

In calculation cases existing at this time, only G1 (1.3 GVA machine) is furnished with AVR/PSS.[16,17]

Figure 8.21 shows the rotor angle swings for the case shown in Figure 8.15 (no-AVR/PSS condition) with and without AVR/PSS.[11,15] The transient is well suppressed and in shorter time the steady state is established. Also, the maximum swing amplitude is much damped. Thus, introducing AVR/PSS, enhancement of stable transmission power limit is possible.

Figure 8.22 shows AVR/PSS variables during the phenomena. During the increase of the rotor angle, by the control of AVR/PSS, the exciting voltage is much enhanced.

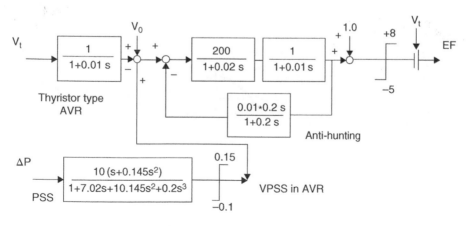

Figure 8.20   AVR/PSS diagram applied to G1.

[16] Data8-111.dat/Draw8-111.acp: Same as the 8-102 case, but high-speed AVR/PSS is applied to the generator G1.
[17] Data8-112.dat/Draw8-112.acp: Same as the 8-101 case, but high-speed AVR/PSS is applied to the generator G1.

Thus, a significant effect of AVR/PSS on the transient stability enhancement is proven.

Next are surveyed a few typical transient stability calculation cases, in which transmission line faulting is followed by clearing of the faulted section. Calculations without AVR/PSS are shown in Figures 8.17 and 8.19. Introducing AVR/PSS,[18] the calculation results are shown in Figures 8.23 and 8.24. In Figure 8.23, the rotor angle swing during the transient is shown, where overloading condition (1.22 GW) is applied.

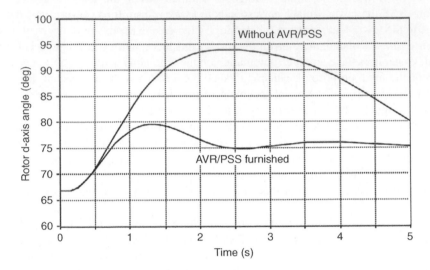

Figure 8.21   Effect of AVR/PSS in one circuit of transmission line opening.

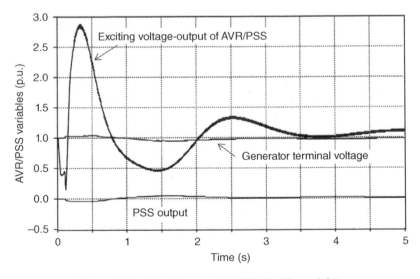

Figure 8.22   Variables in AVR/PSS (for Figure 8.21).

---

[18] Data8-113.dat: Same as 8-103 case, but AVR/PSS is applied.

Even for the overloading, the swing is well suppressed, and thus the stability limit can be enhanced with still some margin by introducing the field excite controlling system. Also, the time duration of the disturbance is significantly shortened. Without such a system, as shown before, by 1.17 GW of transmission power, the system cannot be kept stable. Figure 8.24 shows the AVR/PSS variables during the transient, where PSS output very quickly rises at the first stage. Because of the enhanced the exciting voltage, the rotor swing is strongly suppressed. In this phenomenon, when faced with a great change of the main circuit voltage such as from short-circuiting, AVR/PSS variables show significant performances.

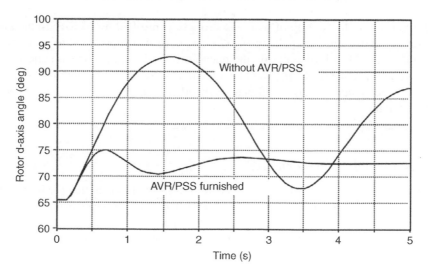

Figure 8.23 3LG—one circuit opening like Figure 8.17 case 1.

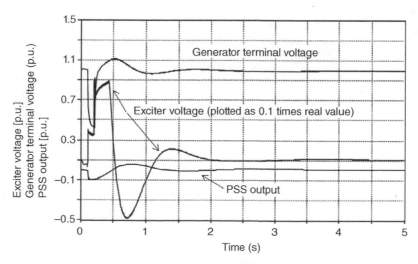

Figure 8.24 AVR/PSS variables in Figure 8.23.

Note:

Great care should be taken in initializing the AVR/PSS variables for accurate calculations. In the EMTP data, initial values of TACS (Transient Analysis of Control Systems) variables are to be directly written. For details, see the data files and *Rule Book*.

In Figure 8.25, two generators versus an infinitive bus case is shown. The system layout is shown in Figure 8.18 where a No. 2 generator is also connected. As shown before, only the No. 1 generator's field exciting is controlled by AVR/PSS.[19] The No. 2 generator's exciting voltage is kept constant during the transient.

The calculated result is shown in Figure 8.25 as for the generators' swing angles. As for G1, the field exciting of which is strongly controlled by AVR/PSS, the swing is naturally and significantly damped. G2's swing is also suppressed to some extent. In Figure 8.26, G2's torque during the phenomenaon is compared with and without field excitation control of G1. The phenomenon is very complicated and the swing period of each machine may be influenced by others. This might be beyond the scope of this chapter.

8.3.4   Back-Swing Phenomenon

Generally, transient short-circuiting current contains a DC component that produces a more or less decelerating direction of torque to the generator. This may influence the transient stability in the relevant power system. The cause of this torque is the trapped flux in the armature winding of the generator during the transient short-circuit current. For details, see Appendix

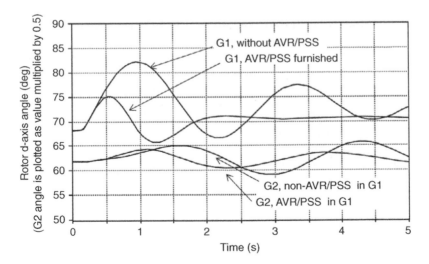

Figure 8.25   Two generators versus infinitive bus case.

---

[19] Data8-114.dat/Draw8-114.acp: Same as 8-105 case, but AVR/PSS is applied to one generator G1.

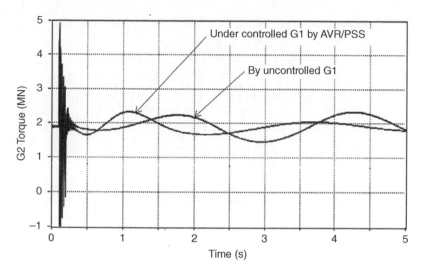

Figure 8.26   G2 electrical torque.

8.A of this chapter. Usually, when applying power frequency phasor domain analysis or the classic graphical method, the phenomena are not represented. The effect of the DC component can only be calculated by applying EMTP.

By the system circuit conditions shown in Figure 8.16 and 8.17, the detailed phenomena are surveyed for both full and less DC components of cases in the fault currents.[20,21]

In Figure 8.27 fault currents and the generator's air gap torque for both full and less DC components are shown. The magnitude of DC component depends on the point on the voltage wave-shape timing of fault initiation. A significant difference between the two conditions is shown in Figure 8.27c. Although the difference in the swing angle is not as significant as that shown in Figure 8.28, a full DC component corresponds to a less severe condition regarding transient stability under such a heavy loading condition as in the present case.

Under a very light loading condition (less power output), a far more significant difference in torque is observed, as shown in Figure 8.29.[22,23] Then, as shown in Figure 8.30, even a negative direction of swing occurs. This is called a back swing. In a very light loading condition, a negative out-of-phase direction might occur. Such phenomena can only be calculated by time domain analysis such as EMTP, as shown here.

More severe conditions exist in pumping stations, where the machine works as a motor. The mechanical load torque is in the opposite direction of the electrical back-swing torque. So, under a heavy load of the pumping-up condition, a DC component in the fault current may introduce an extremely severe condition.

---

[20] Data8-121.dat/Draw8-121.acp: Like the 8-103 case, but maximum of DC component in the fault current condition introduced.

[21] Data8-122.dat: Similar to 8-121, but minimum DC component in the fault current condition introduced.

[22] Data8-131.dat/Draw8-131.acp: Like the 8-121 condition but under very light loading.

[23] Data8-132.dat: Like the 8-122 condition but under very light loading.

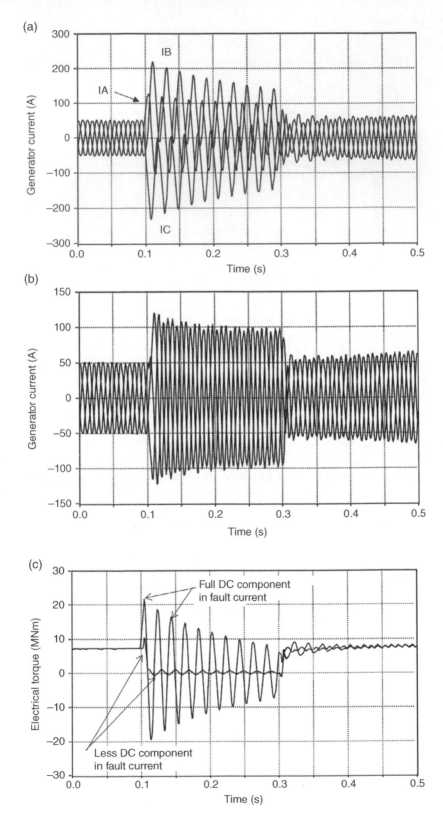

Figure 8.27 Fault current with/without DC component. (a) Fault current with full DC. (b) Fault current with less DC. (c) Torque under full/less DC.

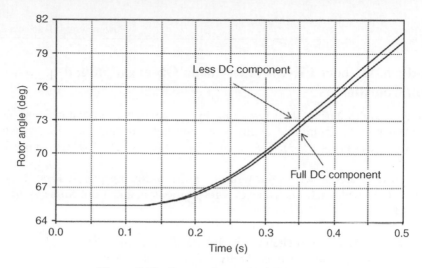

Figure 8.28 Rotor swing angle (full loading).

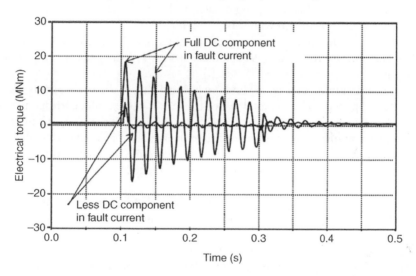

Figure 8.29 Torque under light loading.

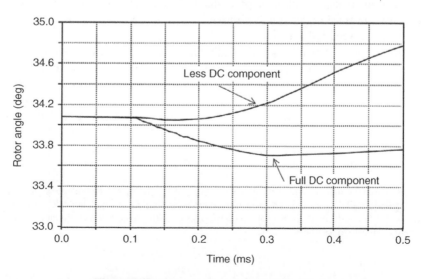

Figure 8.30 Rotor swing angle (light loading).

Appendix 8.A: Short-Circuit Phenomena Observation in d-q Domain Coordinate

Observing variables during short-circuiting in d-q domain coordinates, some new and fruitful ideas may be found that might be a good help for understanding synchronous machine (SM) dynamics [2, 3]. SM rated 3.3 kV, 1 MVA, 2P, 50 Hz is applied as the specimen.[24]

Various results are shown in the figures on the following pages. First, a perfectly round rotor type with each one rotor coil for d/q axis is taken up. Some results are in Figure A8.1.

(a) Three-phase sudden short-circuit currents are shown, where in one phase a full DC component appears; this is termed a fully asymmetrical current.
(b) By short-circuiting a three-phase armature coil, the flux is trapped at the original state with gradual damping. In the d-q coordinate, the basis of which is the rotor position, the flux trace is whirlpool shaped with some damping.
(c) The flux in the rotor, due to the short-circuited coils in both d/q axes, is trapped with gradual damping. Of course q-axis component is kept at approximately zero, which is the origin value.
(d) Armature current trace appears in similar whirlpool shape, where the radius corresponds to the DC component and the rotor distance from (0,0) to the center corresponds to the AC amplitude. The current compensates for the rotor influence on keeping the trapped flux.
(e) The rotor current trace appears symmetrical to the armature one. The original (start) point value corresponds to the steady-state field exiting current. The current is understood to be compensating for the influence from the armature trapped flux.

Next, a symmetrical short-circuit current case is introduced.[25] It should be noted that in the conventional phasor analysis of the case, only this condition can be considered.

(f) Two phases, at the peak-to-peak voltage maximum, are short-circuited. After 90° (1/4 cycle), the rest phase is short-circuited. Then three-phase symmetrical (with no DC component) currents appear.
(g) The specific future in the armature trapped flux. In the time interval of two-step short-circuiting (1/4 cycle), the trace draws a half circle and then afterwards is kept to an almost zero value, that is, no trapped flux.
(h) The effect appears typically in the torque during short-circuiting. Two cases, with and without a DC component in the short circuit current, are shown. The difference is with and without a trapped flux in the armature, while the trapped flux exists in the rotor. It should be noted that in an asymmetrical short-circuit current case that is more usual, though the average of torque is zero, the integration is not zero and may affect the machine velocity change.

---

[24] Data8-11.dat: Perfect round rotor SM without damper coil, 3.3 kV, 1 MVA, 2 P, three-phase simultaneous sudden short-circuiting under no-load condition. (Asymmetrical short-circuit current, i.e., with high DC component).

[25] Data8-12.dat: Similar to the previous, but point on wave short-circuiting to create short-circuit current without the DC component (symmetrical short-circuit current).

Next, short-circuiting under an on-loading condition is taken up.[26]

(i) Armature trapped flux appears initially just rotated due to the load current.
(j) Rotor flux also appears initially rotated.

The next case is for a salient pole machine.[27]

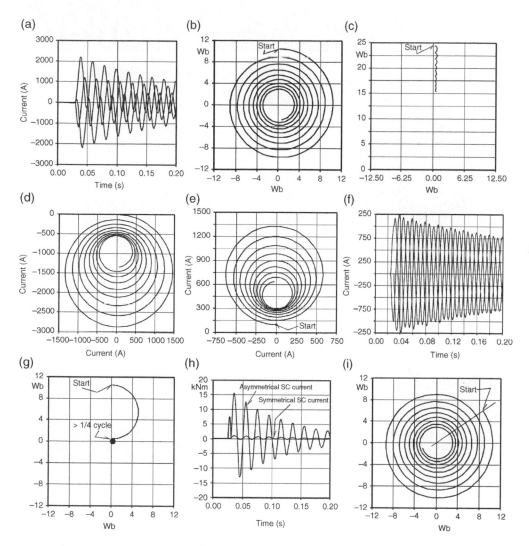

Figure A8.1 Short-circuit (SC) phenomena in the d-q domain coordinate. (a) SC current (asymmetrical). (b) Armature trapped flux. (c) Rotor trapped flux. (d) Armature current. (e) Rotor current. (f) Symmetrical SC current. (g) Armature trapped flux. (h) Torque (asymmetrical/symmetrical current). (i) Armature trapped flux (on-load).

---

[26] Data8-13.dat: Like data case 8-11, but under the full loading condition.
[27] Data8-14.dat: Like data case 8-11, but salient pole machine under a no-load condition.

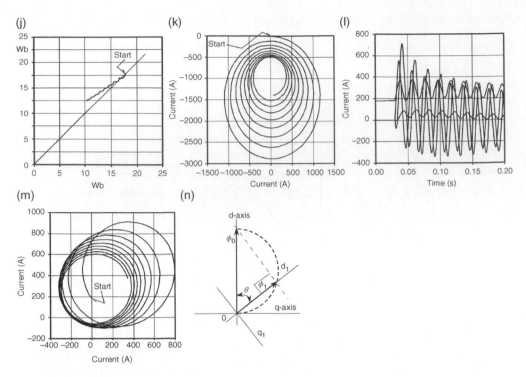

Figure A8.1 (*Continued*) (j) Rotor flux (loaded). (k) Salient pole machine armature current. (l) Four-coil rotor currents. (m) Sum of rotor coil currents. (n) Trapped flux in a symmetrical SC current case.

(k) As for the flux, no significant differences between salient and round rotor ones is observed. But, as for the armature current trace, an oval shaped one appears due to the difference in d/q axis inductances.

Finally, a "four-coil-rotor machine," such as that in Figure 8.1, is taken up.[28]

(l) The armature's variables are almost the same as the two coil one previously mentioned. The four-coil rotor's variables show mutually affected complicated phenomena.
(m) The sum of the rotor d-axis currents in two coils and one q-axis is similar to that of a two-coil machine case.

In (n), the phenomenon in (g) is graphically represented.

In the first instance (two phases are short-circuited), only the horizontal direction component of the armature flux is trapped at zero and then rotates. In the original vertical direction, the flux can change by the induction from the rotor, the value of which is proportional to $\cos\theta$. Then a half circle is drawn during the following quarter cycle of time (90°).

---

[28] Data8-1F.dat: SM with damper coils both in d- and q-axes. Three-phase simultaneous short-circuiting under full load condition.

Appendix 8.B: Starting as an Induction Motor

Synchronous machines can start by direct connection to an AC voltage source, where the damper/field coils work as a squirrel cage of an induction motor. Analyzing such phenomena, a step-by-step initializing adjustment is recommended. For each of the following steps, see the attached DAT file for the detailed parameters.

- The synchronous machine, the ratings of which are 3.3 kV, 1 MVA, 2P, and 50 Hz, is the example here. First, calculation is to be made for normal service (no-load) condition.[29]
- Then the machine's operation in very low speed (i.e., 2% of the rated) is to be calculated for the initial starting condition.[30] Please note that the values of the voltage, frequency, and capacity in the relevant synchronous machine (SM) DAT (in Class 3 SM data card) are to be 2% of the normal rated values. By these, the appropriate machine parameters are automatically introduced.
- Finally, superimposing the relevant regal voltage source connection to the above condition, starting from 2% of the rated speed, the calculation is made.[31]

In Figure B8.1 some calculated results are shown.

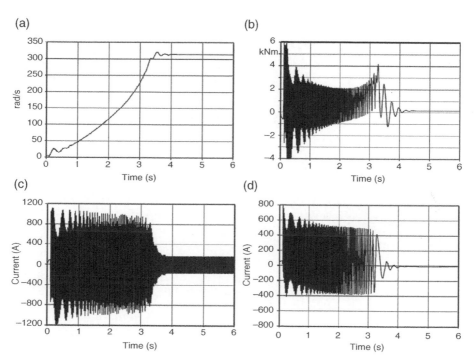

Figure B8.1   Synchronous machine starting in induction machine mode. (a) Velocity (2%–100% of synchronous speed). (b) Torque. (c) Armature current. (d) Rotor d-axis coil current (total).

---

[29] Data8-20.dat: Basic data of synchronous machine starting as an induction machine, operated under power frequency.
[30] Data8-21.dat: Checking operation in very low frequency and very low induced voltage, where 1% of values could be applied.
[31] Data8-22.dat: Starting as induction machine, with 2% of the initial velocity and 1% of the initial voltage, the field current is about 50% of the rating.

(a) Due to the calculation convenience, the initial speed is set to 2%. The machine accelerates from 2 to 100% of the synchronous speed. By the effect of the field magnetizing, the final speed can be 100% irrespective of induction machine mode (slip zero).

(b) In an induction machine, where machine parameters' optimizations seem to be poor, the torque characteristic involves a lot of fluctuations. Yet the machine successfully accelerates to the final (synchronous) speed.

(c) The armature current corresponds to the starting inrush one.

(d) The sum of the d-axis currents (field and damper) is shown. In an SM of EMTP, all coils in the rotor are modeled in unique turn number, so the sum of these corresponds to the d-axis magnetizing. The initial and the final state field current correspond to the field coil magnetizing. The current during the acceleration corresponds to the squirrel cage–type of induction machine.

Appendix 8.C: Modeling by the No. 19 Universal Machine

It is known, for a type No. 58/59 synchronous machine, the number of machines is limited due to an occasional numerical instability. On the other hand, according to the *Rule Book*, applying No. 19 universal machine (UM) modeling, multimachine cases are easily calculated with no limitation in the number of machines.[32,33] Also, by using a No. 19 universal machine, the synchronous machine is applicable, though the calculation condition is less sophisticated compared to Type 58/59 synchronous machines.

As an example, a synchronous machine in Standard UM modeling style is applied in modeling G2 of Figure 8.18, where AVR/PSS is applied to only G1.

For initialization of load flow condition in a certain system with universal machine(s), FIX SOURCES (initializing program) is not suitably applicable. From the author's experience, fixing the generator terminal voltages and (initial) phase angles to all generators based on the infinitive voltage source has been suitably applicable.

Note:

In calculating for a system where all machines are of the No. 58/59 SM type, FIX SOURCES is suitably applicable.

In Figures C8.1 through C8.3, calculated results by two types of SM modeling for G2 are shown where in each of three graphs variables by two types of modeling are drawn superimposed. The difference between the modeling types is not significant, so any kind seems to be applicable depending on the user's choice. Nevertheless, as shown before, application of AVR/PSS and/or power flow initialization is easier and smarter in using Type 58/59 model.

Note:

Calculations of up to eight No. 58 SMs of cases have been successfully made.[34,35]

---

[32] Data8-116.dat: Same as 8–105 case, but one generator is modeled by universal machine model (Type 59 format, or alternatively, synchronous WI in ATPDraw is applicable in No. 19 universal machine modeling).

[33] Data8-117.dat/Draw8-117.acp: Same as 8-105, but the general universal machine's synchronous machine model (Standard UM format).

[34] Data8-115.dat/Draw8-115.acp: Eight generators versus the infinitive bus case, all modeled by a No. 58 synchronous machine.

[35] Data8-1L01.dat: 50 km, double-circuited transmission line parameter calculation, Pi type modeling representation. Reference only.

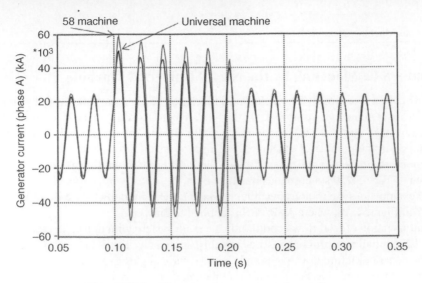

Figure C8.1 G2 current by two types of modeling.

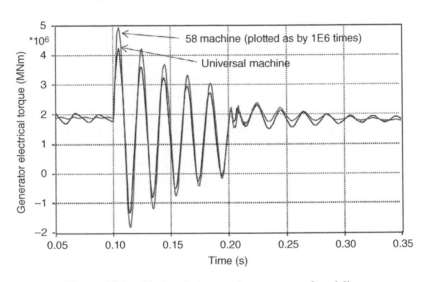

Figure C8.2 G2 electrical torque by two types of modeling.

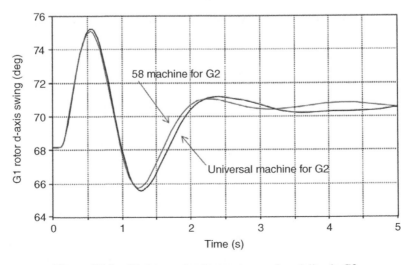

Figure C8.3 G1 rotor swing by two types of modeling in G2.

Appendix 8.D: Example of ATPDraw Picture File: Draw8-111.acp (Figure D8.1).[15]

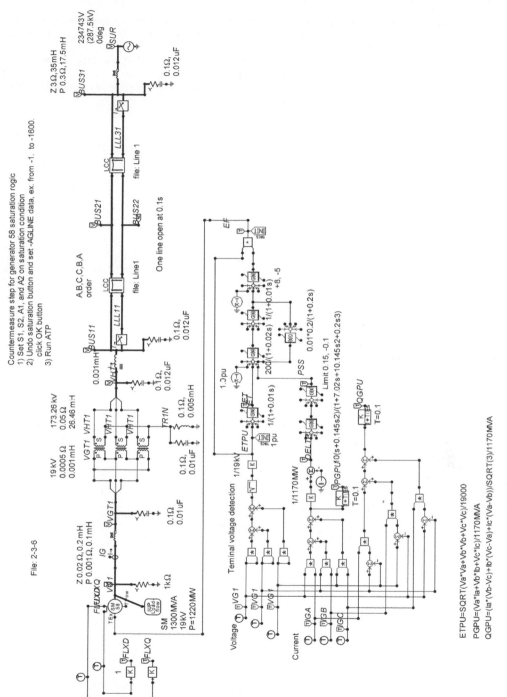

Figure D8.1 2G versus infinitive transient stability analysis, AVR/PSS applied to one generator.

References

[1] X. Cao, A. Kurita, H. Mitsuma, Y. Tada, H. Okamoto (1997) Improvement of numerical stability of electro-magnetic transients simulation by use of phase-domain synchronous machine model, *IEEJ Transactions on Power and Energy*, **117-B** (4), 594–600 (in Japanese).

[2] E. Haginomori, S. Ohtsuka (2002) Surveying synchronous machine sudden short circuiting phenomena by EMTP, *IEEJ Transactions on Power and Energy*, **122-B** (10), 1096–1103 (in Japanese).

[3] E. Haginomori, S. Ohtsuka (2004) Sudden short circuiting of synchronous machines by EMTP, *Electrical Engineering in Japan*, **146** (1), 78–88.

9

Induction Machine, Doubly Fed Machine, Permanent Magnet Machine

According to the "Universal Machines" menu in the *ATP-EMTP Rule Book*, almost all kinds of rotating machines can be analyzed. This chapter deals with machines that are relatively widely used in power systems, industry, or housing facilities, such as the following:

- Cage rotor induction machine (motor), most widely used in general motor usage today.
- Doubly fed machine, the hardware of which is a wound rotor induction machine. The machine is applied to a variable-speed pumping station's motor/generator or flywheel generator.
- Permanent magnet machine. A permanent magnet can be modeled by a fixed value of current source–energized solenoid. In this chapter, the machine is modeled as a synchronous machine in the Universal Machines menu, the field coil of which is energized by a fixed value of current source.

As for a "universal machines" application to synchronous machines, see the previous Chapter 8.

Note:

Presently, ATPDraw is not conveniently applicable to the general application usage of rotating machines. In this chapter, therefore, the usage of ATPDraw is limited.

9.1 Induction Machine (Cage Rotor Type)

Among various kinds of motors, cage rotor type induction machines (motors) are most widely applied to industrial and home facility usage, and even for trains. Owing to the development of power electronics technology, the machine has become applicable even for very quick, dynamic

---

Power System Transient Analysis: Theory and Practice using Simulation Programs (ATP-EMTP), First Edition.
Eiichi Haginomori, Tadashi Koshiduka, Junichi Arai, and Hisatochi Ikeda.
© 2016 John Wiley & Sons, Ltd. Published 2016 by John Wiley & Sons, Ltd.
Companion website: www.wiley.com/go/haginomori_Ikeda/power

responses of operation in a wide range of speeds. The time domain analysis, which is required and applicable to calculate quick dynamic response, is explained in the *Rule Book* as No. 19 Universal Machines. Provided the readers of this text have studied the *Rule Book* well, in this chapter there are some interesting examples peculiar to time domain analyses of the machines.

9.1.1   Machine Data for EMTP Calculation

In the No. 19 Universal Machine menu in the *Rule Book*, machine data, such as motor winding's inductances/resistances, are required. If these data have been obtained beforehand, these can be directly applied to the calculation. Nevertheless, to obtain these quickly is not easy, especially for novice engineers. Today, fortunately, the ATPDraw program is applicable to prepare the data. "Induction WI" is the program menu to calculate the data in ATPDraw.

Note:

In the past, for calculating such machine data, the Windsyn.exe program had been conveniently applied. This program is no longer applicable to today's new (64-bit) PCs. Fortunately, the Induction WI menu in ATPDraw involves a similar function to calculate the data. Therefore, in this chapter, Induction WI is used for calculating the data.

The ACP and ATP files[1] are the example applied in this chapter. First, the following are to be inputted to the Induction WI window in ATPDraw; they are called manufacturer's data or specification/rating data, as shown in Appendix 9.B:

Type		Deep bar cage rotor type	
Frequency	60 Hz	Capacity	500 HP
Speed (synchronous)	1800 rpm	Power factor	0.92
Efficiency	0.96	Slip (rated)	1.7%
Start current	7 p.u.	Start torque	1.2 p.u.
Load torque	1	Max. torque	3.5 p.u.
Cage factor	0.6		

By the process "Fit and View" and, further, "Refit," for fixing the ACP file appropriately, the ATP file is produced, where the following are fixed in the ATP file:

Magnetizing (d/q axis)	0.20091748036 H	
Stator coil (zero sequence)	None	
Ditto of d/q axis	0.5607349643 Ω	0.0090245423515 H
Rotor coil (d/q axis)		
No. 1	0.69390451542 Ω	0.0000000000000 H
No. 2	1.4513984673 Ω	0.0094843215034 H

The deep bar coil is represented by the above two coils.

---

[1] Draw9-01.acp/Draw9-01.atp: Induction machine data calculation and produced ATP file involving the machine data which are applied to cases 2–9 of this chapter.

These data can be directly introduced to the relevant DAT file(s) manually, see the DAT files.

The rotor coil turn number is assumed to be the same as the stator one. Therefore, to obtain the actual rotor current, a calculation depending on the turn ratio is to be made.

In the magnetizing branch, a saturation characteristic can be introduced. For details, see the *Rule Book*.

If a saturation characteristic is also required for stator winding, a saturable reactor is to be connected outside of the motor. Generally, applying inductance with a saturated value can produce an appropriate result.

For the stator zero sequence parameter, in general, nothing is required to be written.

Note:

Before these data are inputted, the relevant chapter(s) of the *Rule Book* and also the relevant HELP in ATPDraw are to be carefully studied.

It should be also noted that the content of the process of "Fit and View" and "Refit" in fixing the ACP file, in which some modifications may be made, does not seem to be recorded in the ACP file. Therefore, for repeating the same calculation, the relevant ACP file is not always directly applicable. It is most essential to apply the relevant ATP file, values of which are to be introduced to the relevant DAT file.

At the mechanical output terminal, a mechanical system can be connected, where it is represented by electric variables/circuit elements in the following way:

Torque (N-m)	Current (A)
Velocity (rad/s)	Voltage (V)
Viscosity (N-m/rad/s)	Resistance (Ω)
Moment of inertia (kg-m^2)	Capacitance (F)

Other variables/elements could be represented by transient analysis of control systems (TACS) controlled sources/EL elements.

9.1.2 Zero Starting

Starting from the stalled (zero) stage can be simply calculated by connecting a voltage source to the relevant motor[2]. But considering a more general application, including plural motor cases, where one unique initialization method for all of the machines is required, a more universally applicable initialization method should be applied. For simple initialization, specifying a slip value for each machine's initial condition seems to be the most convenient application. Therefore, a stalled motor is best represented by connection to the relevant source via a high ohmic resistor and specifying 100% of slip value. Some typical calculated results are shown in Figure 9.1. For details

---

[2] Data9-01.dat: Induction machine zero-starting, three-phase simultaneous SW-ON, 100% of slip applied.

about the DAT file coding, see the data file.[2] The results in Figure 9.1, some of which have been introduced for fixing the ACP file (specifications, rated values, etc.), can be checked.

Some observations about the results are:

(a) The machine accelerates regularly, as expected, to the quasi-synchronous speed.
(b) The overall torque change seems to be regular, but initially it shows violent fluctuation. As shown in the next example, the course seems to be the DC component in the stator coil current after switching on.

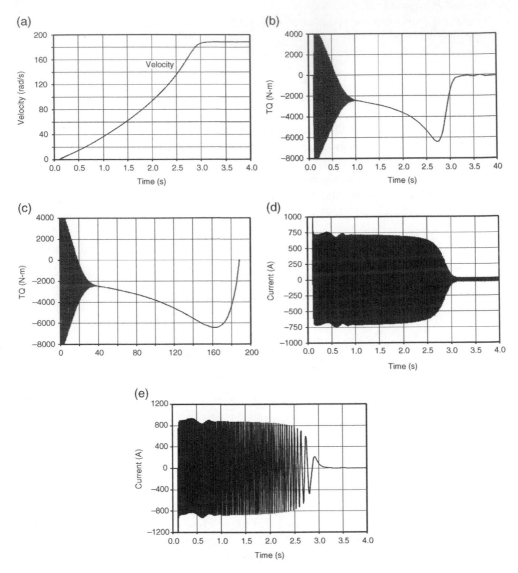

Figure 9.1 Starting from the stalled state—three-phase simultaneous switching. (a) Velocity versus time characteristic. (b) EL (air gap) torque versus time characteristic. (c) EL torque versus velocity characteristic. (d) Stator current. (e) Rotor current (d-axis total).

(c) The popular torque versus speed curve can be made applying the PLOT-XY function of the graphical program. The initial (starting) torque is a little higher than 2000 Nm, which is close to the design value (1.2 p.u. in Fit and View in ATPDraw). The highest torque is approximately 3.5 p.u. as expected, which is generally difficult to design correctly, but in this case, close to the expected value is attained.
(d) The initial armature current is approximately 7 p.u. as expected. After reaching the (quasi-) synchronous speed, the current value is sufficiently low.
(e) Rotor current frequency, on the other hand, depends on the slip value. After reaching close to synchronous speed, the frequency is very low.

As a whole, the motor seems to be well designed (from an electrical point of view) by ATPDraw.

Note:

In this case, for the calculation (solution) process, "compensation" is applied, which is thought to be more accurate and stable for most calculation cases than the "prediction" method. Also, the neutral point of the stator (armature) coil is earthed through a very high ohmic resistor, representing most of the actual cases. Another method may be required in other kinds of calculations, such as plural motor cases.

In Figure 9.1b, c the starting torque shows violent fluctuation. In the stator and rotor currents at the start, a high value of transients appeared (DC component), which was thought to be the course of the torque fluctuation. Therefore, in the next case example,[3] for mitigating the violent fluctuation, producing less of the DC component of starting current was intended. Generally, switching timing dominates the DC component.

It is known that in a mainly inductive circuit, switching on at the source voltage peak timing produces less DC component in the current. In this case, as the stator circuit is actually of an isolated neutral condition (earthed through a very high ohmic resistor), the following process is applied:

- Phase B and C are closed at the P-P voltage peak timing.
- The last phase A is closed 90° later (at the phase voltage peak).

Note:

In this case and the following, the active/reactive/apparent powers were calculated by the following equations, which are correct in the three-phase symmetrical condition.

---

[3] Data9-02.dat: Ditto, but three-phase nonsimultaneous SW-ON for min. DC component in machine current.

For the voltage and current values, instantaneous values are to be applied to the equation.

$$Active\ power = V_a I_a + V_b I_b + V_c I_c \tag{9.1}$$

$$Reactive\ power = \left\{ (V_c - V_b)I_a + (V_a - V_c)I_b + (V_b - V_a)I_c \right\} / \sqrt{3} \tag{9.2}$$

$$Apparent\ power = \sqrt{(Active\ Power)^2 + (Reactive\ Power)^2} \tag{9.3}$$

The calculated results are, in principle, stable.

The calculated results are shown in Figure 9.2, contrasting with Figure 9.1.

Some observations in contrast with the former case are:

(a) The starting up velocity characteristic is almost the same as the former case.
(b) Most significant difference appears in the torque characteristic. Initially, far less fluctuation appears. The average value is almost the same as the former.
(c) Similarly, torque versus velocity curve involves less fluctuation.
(d) The stator coil current involves less initial transient DC components.
(e) Ditto for the rotor coil current.
(f) During acceleration, the power consumption rate is very high for active, reactive, and apparent powers. Due to the high current (reactive/apparent), much active power is consumed by the Joule effect in coil resistances.

9.1.3   Mechanical Torque Load Application

In Figure 9.3a, the steady state (approximately 130% overloaded) calculated result is shown, where, connecting the motor to the source and specifying slip value, the initial state is automatically calculated.[4]

In Figure 9.3b, mainly under noload operation conditions, suddenly mechanical torque, such as mechanical breaking, is applied, which is made, in the DAT file, by specifying the current source value representing the torque.[5] The time lag in the velocity change depends on the moment of inertia of the motor, resulting in time lag in the slip value change and torque increase.

Calculated active/reactive/apparent powers are shown in Figure 9.3c, where relatively high power factor is kept during the phenomena.

Using a friction clutch, a mechanical load is rather smoothly coupled with the motor.[6] Clutch and load are represented by electrical (EL) circuits, as shown in Figure 9.4a.

---

[4] Data9-03.dat: Steady-state loaded condition.
[5] Data9-04.dat: No-load running and then sudden mechanical load torque is applied.
[6] Data9-05.dat: Mechanical load starting via friction clutch connection to the running motor.

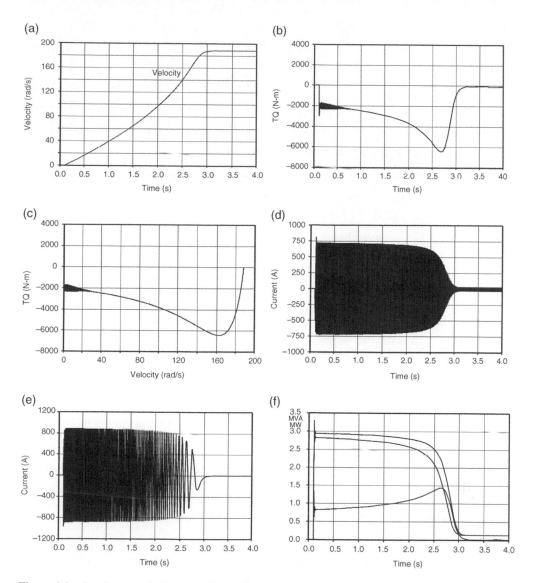

Figure 9.2 Starting up with less transient DC component by point on the voltage wave switching on. (a) Velocity versus time (starting up). (b) Torque versus time at starting. (c) Torque versus velocity characteristic. (d) Stator (armature) current. (e) Rotor current. (f) Active/reactive/apparent powers.

(a) The friction clutch characteristic is represented by a nonlinear resistor shown in the figure. By switching on the clutch connection, the calculation is made.

(b) As for velocities, those of the motor itself and the load are shown. As for torque, the motor's air gap torque and the driving load torque are shown. The main reason for the difference between the two seems to be the motor's moment of inertia and also mechanical loss.

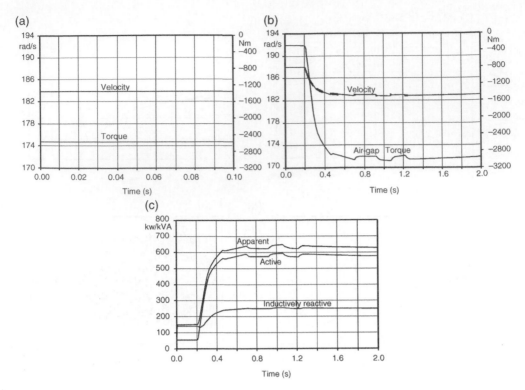

Figure 9.3 Steady state and sudden mechanical torque load application. (a) Steady state, about 130% loading. (b) Sudden mechanical load application (150%). (c) Power consumption during mechanical torque application.

(c) Active/reactive/apparent powers during the coupling are shown. These values depend on the slip increase. During the phenomenon, max slip value change is in the order of 15%, very high, so significantly high values of power change appear.

(d) Comparing EL and mechanical powers, the following conclusions are drawn:

The difference between the EL active power consumption and the total mechanical output is due to accelerating the motor itself and the mechanical/electrical losses of the motor. The difference between the total mechanical output and load driving is due to the friction clutch loss.

9.1.4 Multimachines

The restrictions in multimachine cases[7] are as follows:

- Uniform initialization mode is to be applied to all of the machines in a No. 19 universal machine menu's calculation.
- In initialization, specifying slip to every machine produces the most stable result in the author's experience.

---

[7] Data9-06.dat: Tow machines in a system, one in zero-starting, the other running heavily loaded.

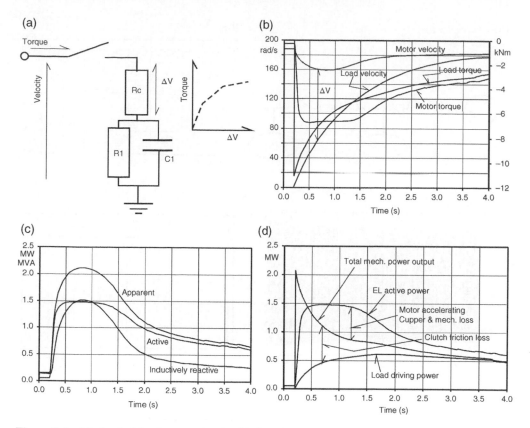

Figure 9.4 Mechanical load connection by friction clutch. (a) Friction clutch and load in EL circuit. (b) Velocity and torque when clutch connecting. (c) Various EL powers. (d) Mechanical and EL various (active) powers.

- "PREDICTION" in the solution and solidly earthed armature coil neutral are mandatory for multimachine cases.
- For cases where there are more than three machines (generally), the "ABSOLUTE U.M. DIMENSIONS" card is to be inserted. Details are shown in the *Rule Book*.

In Figure 9.5 some calculated results for two machines are shown, where

(a) In the system, where M1 is under loaded (120% of the rated) operation condition, M2 starts from a zero stalled state. As shown before, for M1 the relevant slip is specified, while M2 is connected to the system via a high ohmic resistor and 100% of slip is specified, representing the stalled condition. The high ohmic resistor is bypassed at starting.
(b) When M2 starts, due to the high starting current, M1 terminal voltage drops a little; therefore, M1's velocity drops a little.
(c) M1's terminal voltage drops during M2 starting, due to high current consumption in M2.
(d) By the voltage drop, in order to keep the torque constant as specified, except for the transient time interval, the motor-consuming active/reactive/apparent powers are kept almost constant—for example, motor current increases.

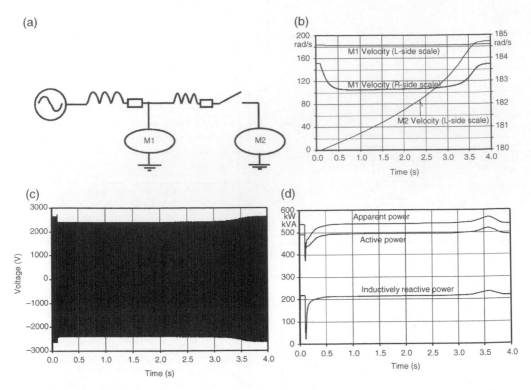

Figure 9.5 M2 starting under M1 loaded steady-state operation. (a) Layout with two machines. (b) Velocities of M1 and M2. (c) M1 terminal voltage. (d) M1 active/reactive/apparent powers.

9.1.5   Motor Terminal Voltage Change

Introducing a simple calculation, where constant torque value is specified and the source voltage is changed[8], the results are shown in Figure 9.6.

(a) The source voltage is intentionally descended in two steps, by 10% in each. To keep the specified constant torque value, the current is enhanced in approximately inverse proportion to the voltage.

(b) The torque value is kept constant as specified, except for the voltage change transient time interval. The velocity goes down a little bit by the voltage drop.

(c) The most interesting thing is active/reactive/apparent powers are kept quasi-constant by the voltage change except for the voltage change transient time interval, that is, the induction motor is considered a constant power consumption load. As shown in Figure 9.6a, the load current increases by the voltage drop. This phenomenon is important from a voltage stability point of view in power systems.

---

[8] Data9-07.dat: Source voltage descended under the heavily loaded running condition.

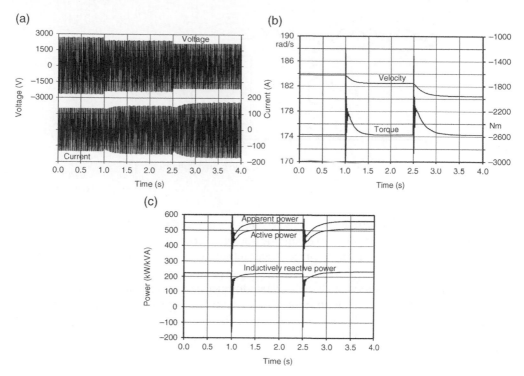

Figure 9.6 Motor terminal voltage change. (a) Motor voltage/current. (b) Velocity and torque. (c) Various powers.

9.1.6 Driving by Variable Voltage and Frequency Source (VVVF)

A cage rotor type induction motor is represented by the equivalent EL circuit shown in Figure 9.7. In the circuit, the motor current I_1 and torque T are calculated as in the following equations[9]:

$$|I_1| = \frac{E_0}{\omega M} \sqrt{\frac{(R_2/\omega s)^2 + (M+L_2)^2}{(R_2/\omega s)^2 + L_2^2}} \qquad T = 3\frac{E_0^2}{\omega^2} \frac{R_2/\omega s}{(R_2/\omega s)^2 + L_2^2} \qquad (9.4)$$

In the equations, if E_0/ω (voltage by frequency) and ω_s (slip frequency value) are assumed to be constant, both the current and the torque are fixed provided variable amplitude/frequency of voltage source is applied. Thus, any kind of control can be made.

Practically as for the motor voltage, V_1 is to be applied instead of E_0. There is a small difference between V_1 and E_0, as shown on the right-hand side of Figure 9.7. In some cases, the difference is to be considered for very accurate control.

[9] Data9-08.dat: VVVF (linearly rising voltage and frequency) source starting.

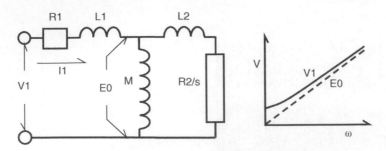

Figure 9.7   Induction machine equivalent circuit.

As a variable amplitude/frequency voltage source, a power electronics (inverter) is applied today. But in this chapter on surveying the motor's basic principle, a TACS-controlled (idealized) source in the Electromagnetic Transients Program (EMTP) is applied. (Also see Chapter 6 of this book.)

In the example case introduced here, constant torque/acceleration rate starting is taken up in comparison with the direct starting shown in Figure 9.2. In the case where E_0/ω is kept constant, ω is enhanced at a constant rate. Thus, the constant torque is obtained (see the attached DAT file for details). The result is shown in Figure 9.8.

(a) The velocity change is, as intended, similar to the direct starting of case. The torque is also, as intended, mostly constant during the acceleration time interval.
(b) The most significant difference between the two types of starting is EL power consumptions. In direct starting, very high mainly reactive current flows, especially at the initial low velocity region. On the other hand, by variable amplitude/frequency voltage source, EL power consumption is far lower. The very high active power consumption by direct starting seems to be due to copper loss in the coils from the very high current. Thus, for frequent start/stop of application case, VVVF produces apparent merit.
(c) EL active power consumption and the air gap torque output power (mechanical) are compared. The difference depends on the EL side loss only, that is, copper loss in the rotor, since no iron loss is applied in the calculation model. In any case, the efficiency is very high.
(d) For the purpose of less violence especially initially, source voltage frequency change is mitigated. For details, see the TACS part of the DAT file.
(e) The source voltage wave shape is shown, where the amplitude and frequency rise linearly. This called a TACS-controlled voltage source. It should be noted that in TACS, voltage itself is to be represented by a formula. The relevant voltage is to be written as $V\cos\theta$ instead of $V\cos\omega t$, where θ is the angle at the relevant time, that is, time integration of ωt.
(f) Stator and rotor currents are shown. The frequency of the rotor current is equal to the slip frequency, which is specified to be mostly constant. The figure shows this condition.

As a whole, optimized starting is established.

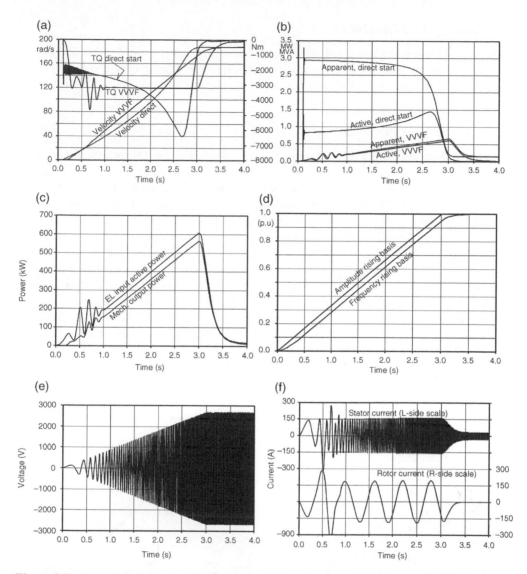

Figure 9.8 Starting with constant torque by a variable voltage, variable frequency (VVVF) source. (a) Velocities and torques at starting. (b) Power consumptions at starting. (c) Power consumption details by VVVF. (d) Frequency change mitigation. (e) Source voltage wave shape. (f) Stator and rotor currents.

9.2   Doubly Fed Machine

The physical construction of a doubly fed machine is, in principle, identical to one of a wound rotor type induction machine[10-13]. The machine is applicable to a variable-speed generator/motor and, furthermore, to a flywheel generator or synchronous compensator.

9.2.1   Operation Principle

The rotor coil is, via a slip ring, energized by the outside source. The principle of operation is shown in Figure 9.9. In (a) the image of the machine's physical construction is shown, and in (b) the phasor vector diagram is shown. Please note, the vector angles of the secondary side variables are based on the primary side. The actual secondary side injected voltage/current is to be based on the rotor position. Details will be explained later. Also, the turn number of the secondary (rotor) coil is assumed to be the same as the primary side. X1 and X2 are the leakage reactance of the primary side and secondary side, respectively, where both are based on the synchronous frequency. "s" is the slip value; for example, in overspeed condition, s is negative. "ϕ" is the common flux linkage and I0 is the exciting current.

V1 is the system (terminal) voltage and is generally fixed. If any of V2 or I2 is given, the total vector diagram is completed, that is, by V2 or I2 the target operation condition is attained. Actually for relatively quick change of controlling, I2 is superior than V2 regarding the stable operation of controlling. As shown in the appendix, the machine is, in principle, applicable to any kind of generator, motor, compensator, or flywheel operation.

Please note in the figure, the resistance in the stator/rotor coil and exciting loss are excluded. If a very accurate calculation is required, these are to be considered.

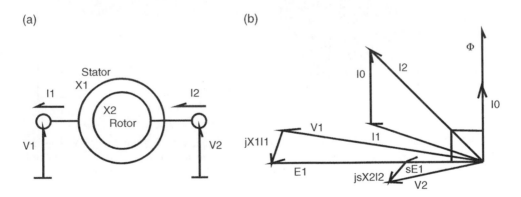

Figure 9.9   Doubly fed machine construction and operation principle. (a) Physical construction. (b) Phasor vector—primary side view.

---

[10] Data9-11.dat: Doubly fed machine steady state, secondary side is fed by voltage source.

[11] Data9-12.dat: Same as 9-11, but secondary side is fed by current source.

[12] Data9-15.dat: Doubly fed machine flywheel generator operation.

[13] Draw9-02.acp/Draw9-02.atp: Example of doubly fed machine data calculation and produced ATP file. For reference only.

In the following data cases, as targeting a flywheel generator in a power system for the enhancement of the system dynamic stability, the following machine is applied:

Ratings as a wound rotor induction machine			
Voltage	24 kV	Capacity	100 MW
Frequency	50 Hz	Synchronous speed	500 r.p.m.
Inertia constant	10 s	Power factor	0.9
Load efficiency	0.97		

Machine data used in No. 19 Universal Machine calculation	
Magnetizing inductance	0.088234 H (d/q axis)
Stator leakage inductance	0.001617 H (ditto)
Ditto resistance	0.087169 Ω
Rotor leakage inductance	0.001617 H
Ditto resistance	0.127258 Ω

Please note, to calculate the above machine data, another program (Windsyn.exe) was used. Today, alternatively, "Induction WI" in ATPDraw is applicable, which produces similar results. (See, as an example, the files.[13])

9.2.2 Steady-State Calculation

In Figure 9.10, two kinds of calculation results are shown, in one of which the voltage source is connected to the secondary (rotor) side; in the other case, the current source is connected. For details see the DAT files. For both cases, the calculated results are stable and appropriate. Therefore, in calculating steady state, any voltage or current source can be connected on the secondary side[10,11]. Please note that from two cases the results are not perfectly equal, as the tuning is not quite perfect.

In both cases the machine supplies mainly lagging reactive currents, that is, capacitor bank mode operation.

9.2.3 Flywheel Generator Operation

As the most specific (peculiar) application of the machine, a flywheel generator is taken up[12]. As shown in the Appendix, the machine can be applied for any mode of generator/motor. In the following example case, the machine operates as a flywheel generator, where it creates power by reducing the rotating speed and transferring the rotating energy to the EL power. The machine speed goes down significantly.

As mentioned before, the machine can be controlled by injecting the current to the secondary (rotor) side, as shown in the vector diagram in Figure 9.9. The actual injected current is to be transferred to one based on the secondary side rotor position. For this purpose the rotor position (angle) must be known. The position sensor, located on the rotor, is one possibility. But for a more sophisticated method, sensorless position detecting is desirable here. The principle shown in the following is applicable:

• Measuring the present primary side voltage and current.
• By these, the secondary side current (primary side based) is calculated.

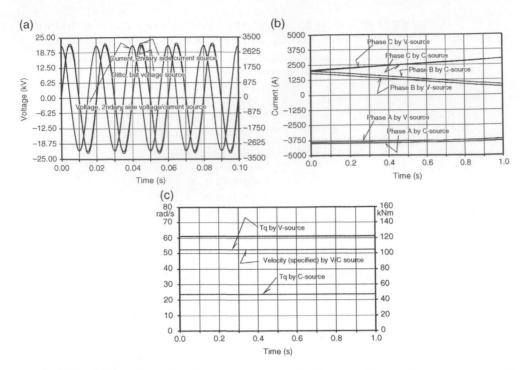

Figure 9.10 Calculated results of the steady state. Results by voltage and current sources are superimposed. (a) Stator voltage/current. (b) Rotor voltage/current. (c) Velocity and air gap torque.

- Measuring the present secondary side current (secondary side based).
- Comparing both present angles, the present rotor position angle is known.

While from the targeted output primary side current (target output power) the relevant (primary side based) secondary side current is calculated, which is to be transformed to one based on the rotor position applying this process, the actual secondary side injection current is obtained.

Figure 9.11 shows the flow chart of the process of transformation.

For details of the program coding of the calculation in TACS, see the data file.

In the example case of flywheel generator operation, the machine is initially mainly in noload operation with 110% of synchronous speed. Then, approximately 150% of overload output power (generator mode) is generated. Consuming the rotating energy, the machine decelerates. It should be noted that the magnitude of the current in the mechanical network, corresponding to the torque, is in the order of MNm. So the connecting resistance in the circuit corresponding to the mechanical viscosity is to be appropriately low, otherwise abnormal velocity violence may be created. Active, reactive, and apparent powers of both primary and secondary sides are calculated and shown. It should also be noted that the secondary side power depends on the slip velocity, that is, mostly in proportion to the slip values. In the calculation, the secondary power source is a TACS-controlled current source. See Figure 9.12 for details. As a whole, as shown in Figure 9.12e,f, the secondary side is to

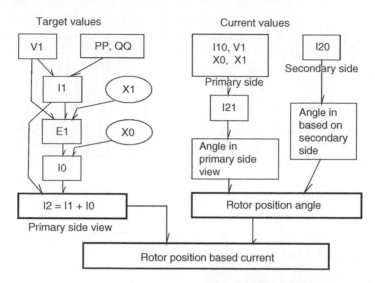

Figure 9.11 Rotor position angle calculation flow chart.

be supplied with power in proportion to the slip value for a constant power supply in the primary side. A higher value of mechanical moment of inertia of the machine corresponds to a lower rate of velocity change, resulting in a lower value of the secondary side's necessary power source capacity.

Some further calculated results are shown in Figure 9.12.

(a) The primary side voltage and current corresponding to the generator output power. At the start of the controlling, current is enhanced.
(b) Secondary side output currents are equal to the reverse of the injected (control) current. The frequency is equal to the slip frequency, so at the middle timing it reversed.
(c) Rotor position angle calculated along the flow chart.
(d) Machine velocity and air-gap torque.
(e) Primary side output powers. Constant active power is generated.
(f) Secondary side injected powers. For 0 > s (lower speed condition), active power is to be injected to the secondary side, leading to power consumption at the secondary side.

9.3 Permanent Magnet Machine

Due to higher efficiency in operation, the permanent magnet machine[14,15] is being put into practical use more often.

A permanent magnet is approximately modeled by a fixed value of current source–energizing reactor. Therefore, a permanent magnet machine is represented by a synchronous machine, the field coil of which is energized by a fixed value of current source.

---

[14] Data9-21.dat: Permanent magnet machine steady-state operation.
[15] Data9-22.dat: Permanent magnet machine, steady-state initializing and then transient operation.

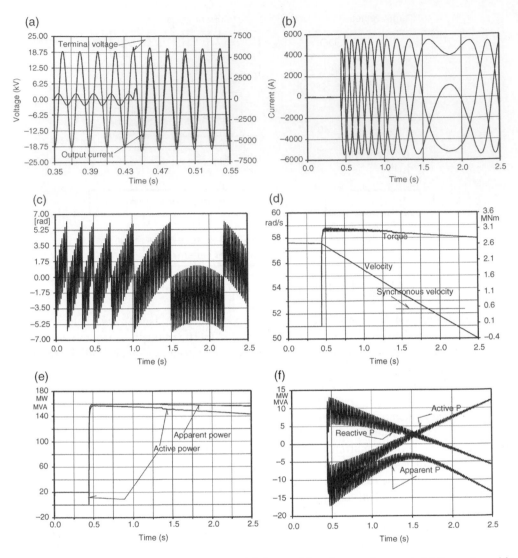

Figure 9.12 Flywheel generator operation. (a) Primary side voltage and current. (b) Secondary side current, three-phases. (c) Rotor position angle. (d) Velocity and air-gap torque. (e) Primary side output powers. (f) Secondary side injecting powers.

Introducing the relevant synchronous machine parameters, the following are applicable:

• Damper coil condition is to be appropriately applied.
• For an interior magnet–type machine, generally the Xd value is lower than Xq.

From the author's experience, Synchronous Machine model in the No. 19 Universal Machine is most conveniently applicable. Introducing certain corresponding synchronous machine data at the first step of the calculation, the appropriate field exiting current value is to be fixed, which can be obtained by the general synchronous machine's initialization process. Then the field

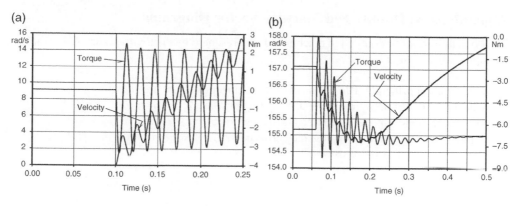

Figure 9.13 Two examples of permanent magnet motor transients. (a) Zero starting as IM (**induction motor**). (b) Voltage drop down by 30%.

energizing source is switched over to the relevant value of current source. After these, calculation of transient phenomenon is to be made. Please note that the relatively high ohmic resistor is to be connected in parallel to the current source, the value of which is to be fixed by trial and error, that is, the true current source is not applicable due to the EMTP calculation algorithm.

Two example cases are shown in Figure 9.13.

9.3.1 Zero Starting (Starting by Direct AC Voltage Source Connection)

The machine with damper winding (equivalent) can start like an induction machine. The field exiting current value is calculated beforehand and applied in the initial condition. In this case, the field current value is sufficiently lower than the generally suitable value for the operation under the relevant rated voltage condition. If the current value is higher (strong magnet), the motor does not start. For this type of direct starting machine[14], a weaker magnet force is necessitated. This may be an interesting subject to study. In the case shown in Figure 9.13a, the fluctuation is still very high compared to a general induction machine. The starting efficiency can never be high.

9.3.2 Calculation of Transient Phenomena

In some cases, the following calculation process is conveniently applicable[15]:

- Initializing the corresponding synchronous machine in the target operation condition where, in general, the field coil is energized by the relevant voltage source. The exciting current value is to be recorded.
- Introducing a current source, the value of which is equal to the one obtained previously.
- Switching over the source, exciting the field, from the initial voltage source to the current source, where the appropriate resistor is to be connected in parallel to the current source as shown before.

Afterward, any calculation of transient as a permanent magnet machine can be made. In the case shown in Figure 9.13b, after establishing the permanent magnet motor initial condition, the source voltage descended by approximately 30% and velocity and torque fluctuations were calculated.

For the practical coding of the calculation data, see the data files.

Appendix 9.A: Doubly Fed Machine Vector Diagrams

The operation states of doubly fed machines are represented by vector diagrams. A few typical examples are shown in Figure A9.1.

(a) Construction principle. Directions of voltage/current vectors are shown.
(b) Generator mode supplying active and lagging power.
(c) Generator mode supplying leading and active power.
(d) Motor mode consuming lagging and active power.

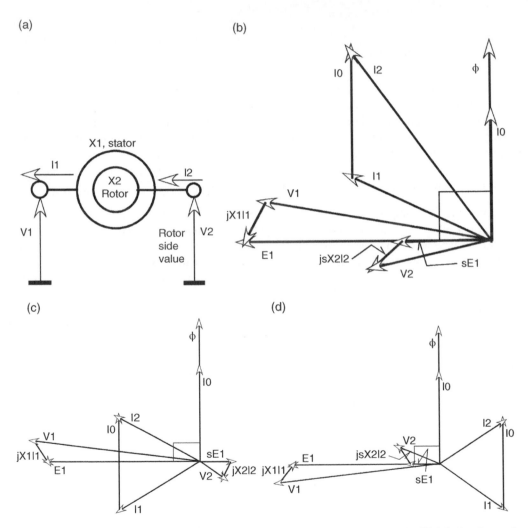

Figure A9.1 Doubly fed machine vector diagrams. (a) Construction principle. (b) Slightly lagging and active power supplying. (c) Leading and active power supplying. (d) Lagging active power consuming (motor).

Appendix 9.B: Example of ATPDraw Picture

ATPDraw file: DRAW9-01.acp[14] Calculating induction machine parameters.

After correctly completing the following windows, the relevant appropriate ATP file can be produced (Figure B9.1).

(a)

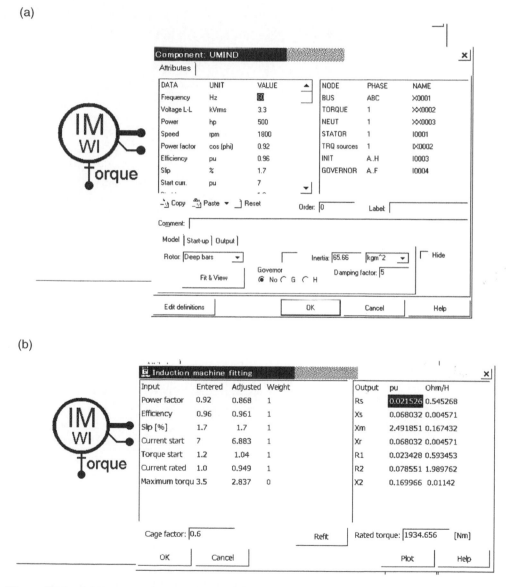

(b)

Figure B9.1 Induction machine data inputting and fitting windows. (a) Machine ratings, and so on inputting window. (b) Machine parameters fitting window.

10

Machine Drive Applications

10.1 Small-Scale System Composed of a Synchronous Generator and Induction Motor

In power systems significantly influenced by rotating machine dynamics, time domain dynamics analysis is of great concern. Today, typical small-scale power systems, such as those of independent power producers, involve synchronous generators for supply and induction motors as a significant part of the load. This chapter will discuss this type of system. A simplified system involving a synchronous generator and two induction motors is shown in Figure 10.1.

10.1.1 Initialization

First, the initialization technique in ATP-EMTP for such a system is to be established. In Figure 10.1a the basic simplified system layout is shown, where power to two induction motors is supplied by a synchronous generator. Assuming one induction motor is in full loading and the other is in very light loading conditions, the following mode of initialization is applied.[1] An ATPDraw picture of this case is shown in Appendix 10.A.

- As all universal machines are to be initialized in uniform initialization mode, slip conditions are given to the two machines.
- Voltage amplitude and phase angle are given to the synchronous generator terminal.
- No fix source or load flow option is specified, as these are not suitably applicable to a universal machine.

---

[1] Data10-1-1.acp: Initialization of a system composed of one synchronous generator and two cage rotor induction motors.

Power System Transient Analysis: Theory and Practice using Simulation Programs (ATP-EMTP), First Edition.
Eiichi Haginomori, Tadashi Koshiduka, Junichi Arai, and Hisatochi Ikeda.
© 2016 John Wiley & Sons, Ltd. Published 2016 by John Wiley & Sons, Ltd.
Companion website: www.wiley.com/go/haginomori_Ikeda/power

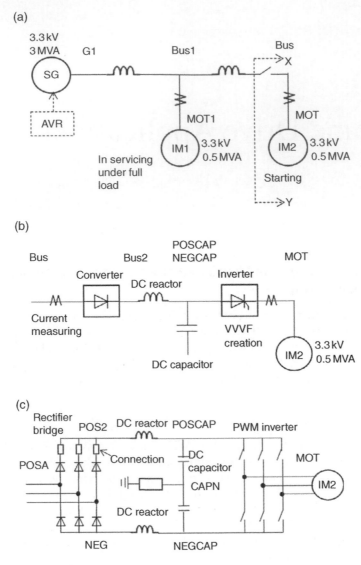

Figure 10.1 Small-size system layout with synchronous machine (SM) and induction motor (IM). (a) Basic small-size system layout. (b) Induction motor VVVF starting. (c) Detail of the VVVF starting circuit layout.

- In Figure 10.1a, the automatic voltage regulator (AVR) is not in service and the switch for IM2 is permanently closed.

Some calculation results are shown in Figure 10.2, where, for convenient comparison, the 2P and 4P machines' variables are shown in common graphs. In (a) each machine's velocity stays constant. In (b) the torques of two machines are also constant. It shows good initialization is obtained.

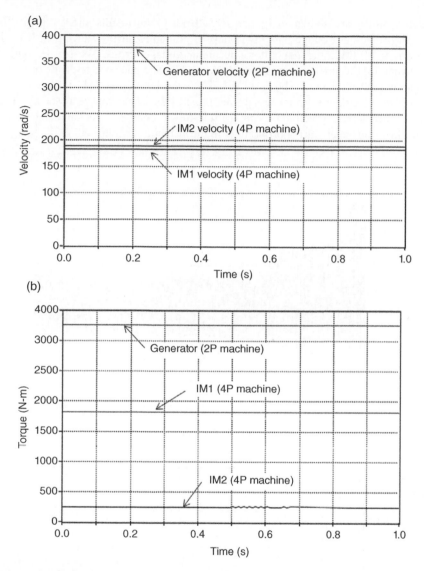

Figure 10.2 Rotating velocities and air gap torques, one synchronous generator (2P), and two induction motors (4P). (a) Rotating velocities. (b) Air gap torques.

10.1.2 Induction Motor Starting

The next trial is the second motor (IM2 in Figure 10.1) starting from the stalled condition. At initialization, in order to represent the "stalled condition" and to satisfy the "uniform initialization mode" condition, the starting switch, which is initially open circuited, is shunted by very high ohmic resistors, that is, very low voltage is applied to the motor, and 100% of initial slip value is given to the motor. Then the starting switch is closed. For details, see the data file.[2]

---

[2] Data10-1-2.acp: Same system as in data file 1, but one of the motors starts directly, resulting in voltage collapse.

Calculated results are shown in Figure 10.3. Briefly, the results show typical "voltage collapse." In (a) the bus voltage is being collapsed gradually during starting of IM2. By the starting current of IM2, the generator's terminal voltage drops, so, for keeping IM1's torque constant, IM1's current increases as shown in (d). The bus voltage drops further. IM1's velocity can never been kept as shown in (b), while acceleration of IM2 is very low. By the voltage drop, the generator supplies less power, that is, less air gap torque as shown in (c), and the generator accelerates gradually as in (b) due to the constant mechanical input torque. As a result, IM2 can never start appropriately in the system.

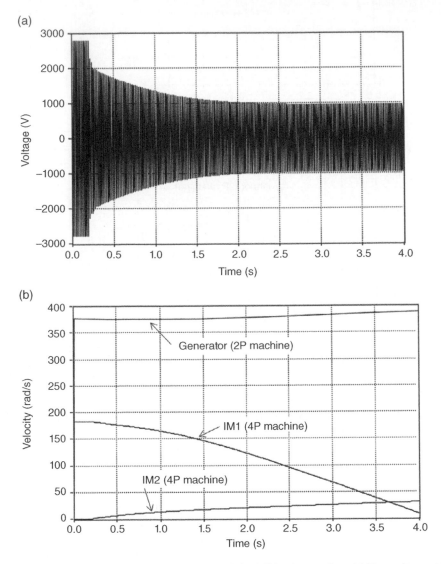

Figure 10.3 One motor is starting while the other is in full-load operation. (a) Bus voltage at BUS1. (b) Generator and motor velocities. (c) Generator and motor torque. (d) IM1 current.

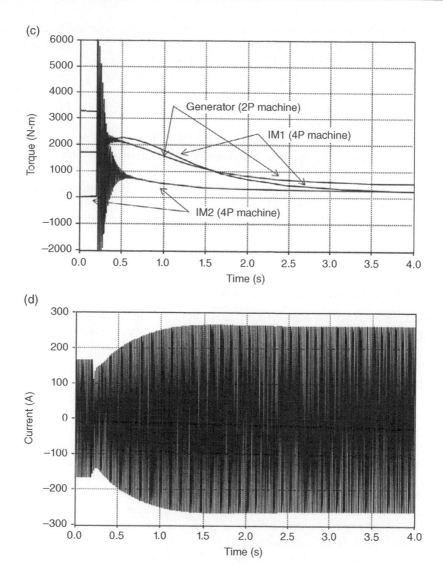

Figure 10.3   (*Continued*)

10.1.3   *Application of AVR*

The most important requirement in this system is to keep the voltage. AVR is the first priority for the purpose. So let us introduce AVR to the generator. In Chapter 8 AVR was discussed. In this chapter, the same AVR (but without a power system stabilizer, PSS) is introduced. Chapter 8 relates to a very high capacity of generators, but for simplification, the same one is applied also to a relatively low capacity of generator in this chapter. For details, see the data file.[3]

---

[3] Data10-1-3.acp: Same system, but the generator is furnished with highly sensitive AVR, resulting in successful starting.

Calculated results are summarized in Figure 10.4a, which shows generator terminal voltage where, though a voltage drop of a short time interval appears at the initial time, the voltage is kept at an approximately constant value. The generator exciting voltage, which is the output of AVR, shows (in b) the initial steep enhancement and the following approximately constant value of 250% of the original one during the motor starting time interval. After the start has been established, the value comes back to approximately the original value. The exciting current (in c) shows the similar variation.

Figure 10.4d shows the motor (IM2) started normally. But the system frequency, that is, the synchronous generator's velocity, lowered a little. The generator air gap torque, shown in (e), due to the enhancement of the field exciting voltage, enhanced a lot during the starting, whereas the mechanical input torque is kept constant due to nongovernor controlling. Therefore, the generator is decelerated.

Figure 10.4f shows that the induction motor consumes extremely high inductively reactive power during starting.

10.1.4   Inverter-Controlled VVVF Starting

As shown in Figure 10.4f, the cage-rotor induction motor's starting consumes extremely high inductively reactive power. While, as shown in Chapter 9, VVVF (variable voltage, variable frequency) starting with linearly rising voltage and frequency provides highly efficient starting. For such purposes, power electronics technology, that is, an inverter, is suitably applicable. The next trial is applying such power electronics technology to the case. Figure 10.1b,c show the circuit layout applied. A pulse width modulation (PWM) inverter is introduced, where practically any kind of AC voltage wave can be produced corresponding to the reference voltage signal wave shape. So, introducing linearly rising amplitude and frequency of wave shape as the reference wave, a suitable VVVF source is obtained in Figure 10.1.

Care should be taken that the PWM inverter shown in Figure 10.1 produces correct voltage for phase-to-phase, but not for phase-to-earth, that is, zero sequence voltage component exists. On the other hand, the induction motor armature coil is to be solidly earthed for automatic initialization due to the restriction in ATP-EMTP. So, undue current of zero sequence component may flow in the armature coil, though the current introduces little effect on the rotation of the machine. Nevertheless, the current may produce undue joule loss. To exclude the zero sequence current, any of the following means can be applied:

- Applying a solidly earthed neutral type inverter.
- Inserting a star-delta connected transformer for infinite zero sequence impedance in the source side.
- Reactor with very high zero sequence impedance inserted.

In Figure 10.1, the third method seems to be the most simple, so this method is applied, though this is not realistic. Figure 10.5 shows calculation results without such considerations. Though the rotation phenomenon seems to be normal, a typical undue armature current with high zero sequence components is the result.

With a significantly high value of zero sequence reactance in the reactor between the source and the rectifier bridge in Figure 10.1, and with high impedance for the capacitor neutral

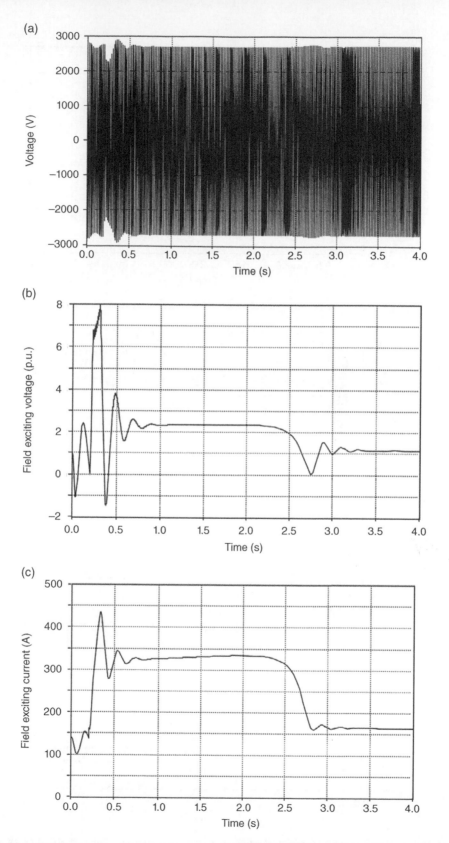

Figure 10.4 Induction motor starting in a system supplied by an AVR-furnished generator. (a) Generator terminal voltage. (b) Generator field exciting voltage—AVR output. (c) Generator field exciting current.

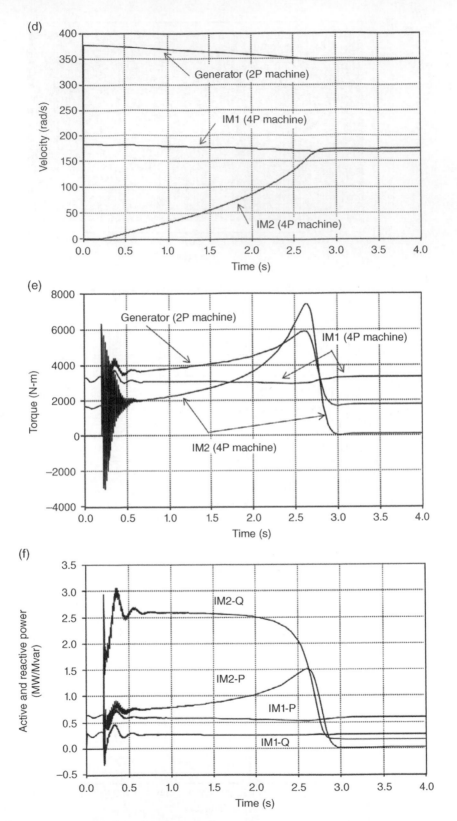

Figure 10.4 (*Continued*) (d) Generator/induction motor velocities. (e) Generator/induction motor torques. (f) Induction motor active and reactive powers.

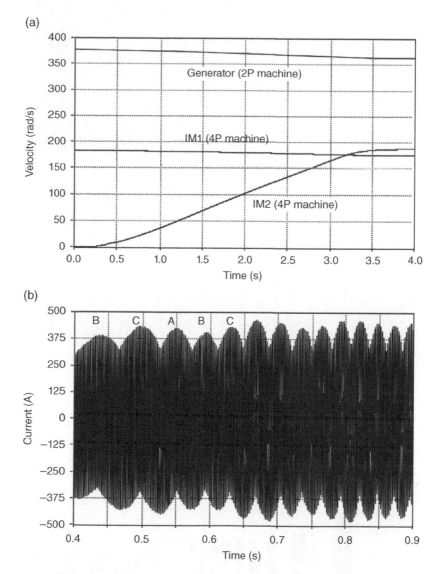

Figure 10.5 Variables in both source and motor earthed system circuit. (a) Velocity change. (b) Motor IM2 armature current.

earthing (at node CAPN), some calculation results are shown in Figure 10.6. For details of the circuit parameters, compare the data files.[4]

The following are noted for each part of Figure 10.6:

(a) The three-phase reference voltage wave shape is shown for the first second, the amplitude and frequency of which rise linearly certain specified values.

---

[4] Data10-1-4.acp: Same system but the starting motor is driven by VVVF inverter source, the neutral potential of which is restricted, resulting in undue zero-sequence motor current.

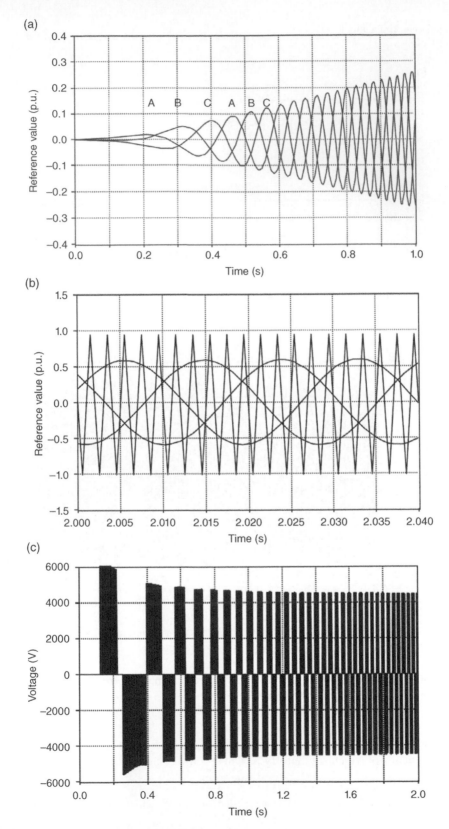

Figure 10.6 Comparison of VVVF beginning with direct starting. (a) Reference voltage wave—initial part. (b) Reference voltage and triangular wave. (c) Motor terminal phase-to-phase voltage.

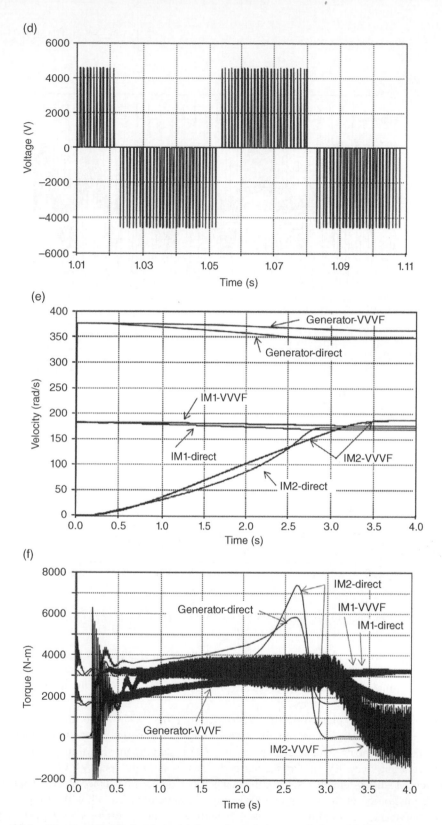

Figure 10.6 (*Continued*) (d) Same as (c) but enlarged time resolution. (e) Generator and motor velocities. (f) Generator and motor torques.

(g)

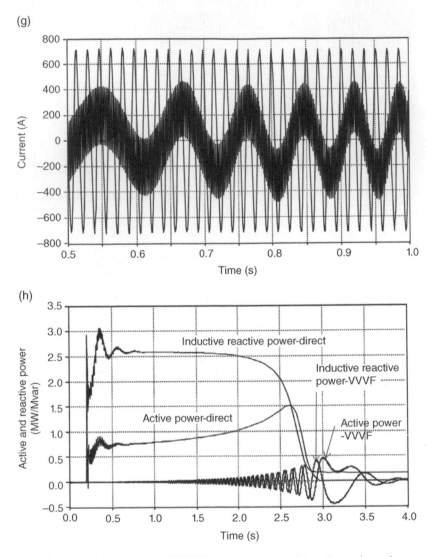

Figure 10.6 (*Continued*) (g) Direct and VVVF starting currents. (h) Active and reactive powers.

(b) Triangular carrier and three-phase reference voltage wave shapes in transient analysis of control systems (TACS) are shown for the intermediate time. By comparing the waves, the gate signals to the inverter valves are produced in TACS for producing correct phase-to-phase voltage to the motor.

(c) The motor terminal phase-to-phase voltage fed by PWM is shown for the first 2 s.

(d) Same as (c), but shown in a very fine time resolution for the intermediate time interval.

(e) Generator and motor velocity changes by the VVVF starting are shown in comparison with direct starting ones. By similar starting performances in both cases, VVVF brings far less influence to the system, that is, less descending in synchronous machine velocity/frequency.

(f) Time integration of the starting motor's torque, that is, the area below the torque-versus-time curve is to be equal for both cases. Nevertheless, the generator's torque curves show a great difference between the two. Significant electrical loss is to be produced, most noticeably by the winding's Joule loss.

(g) Instead of the mostly similar starting characteristics by both, the starting motor currents show great difference from each other. This is the most typical feature of VVVF starting of induction motors: high efficiency.

(h) Active and inductive reactive powers during starting by both direct and VVVF starting are compared in the figure. (For the output of VVVF starting, due to nonsymmetrical three-phase variables creating a lot of noise, the outputs are smoothed in TACS.) Great energy saving in VVVF, especially at starting, is significant. Also, little reactive power is consumed. Within the inverter circuit, reactive power can be produced.

Figure 10.7 shows the generator's field exciting voltages by both direct and VVVF starting; both are controlled by AVR. By VVVF starting, AVR seems to be almost unnecessary. A VVVF-controlled cage-rotor induction motor seems to be applicable for various variable speed controlled usage.

10.2 Cycloconverter

For relatively low frequency of power source such as 10–20 Hz, cycloconverters have been widely applied, the special feature of which is that a high-power and relatively low-price thyristor is applicable as the switching valve element, and the efficiency is high due to direct frequency converting.

In Figure 10.8, a one-phase of cycloconverter circuit is shown, three sets of which compose a three-phase cycloconverter. In a three-phase cycloconverter, a minimum of 36 arms of switching elements are involved, such as in the data file.[5] The .dat file is for EMTP-ATP input file and is not an .acp file for ATPDraw.

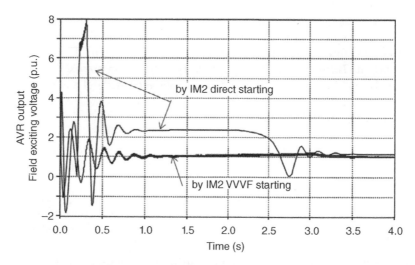

Figure 10.7 Field exciting voltage comparison.

---

[5] ATPData10-2-1.dat: Three-phase cycloconverter circuit, creating 15 Hz of voltage from a 60 Hz source.

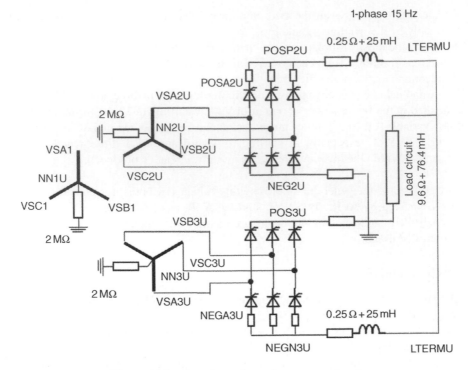

Figure 10.8   One phase of a cycloconverter circuit.

Each phase consists of plus and minus side blocks, and one block is a three-phase thyristor rectifier bridge. In a thyristor converter circuit, as shown before, the output DC voltage is proportional to cosine α, where α is the ignition delay angle; thus, a slow change of α produces a slowly changing DC voltage. Therefore, the upper side of the converter bridge in Figure 10.8 can produce positive polarity of half a wave, and the lower side, negative polarity of one. Consequently, the circuit can produce a relatively low frequency of alternating current. Turning over from one polarity to the other should be smooth. In this case, fortunately, automatic smooth turning over is obtained without any special means, as shown later (Figure 10.9c).

Some calculated results are shown as follows:

Figure 10.9a shows across-bridge voltages of both polarity converter bridges. Each bridge produces one polarity of voltage, but due to the connection to each other, both polarities of voltages are induced on the terminals.

Figure 10.9b shows the Fourier spectrum of the voltage, where relatively high orders of harmonics are involved. Six switches per cycle are performed in the converter bridge, and the output frequency is one-fourth of the system frequency. Therefore, the number of harmonic orders is around $6 \times 4 = 24$.

Figure 10.9c shows both bridges' currents and the load circuit current. Around the load current zero time interval, circulating current through both bridges is observed, producing continuity of the load current at zero.

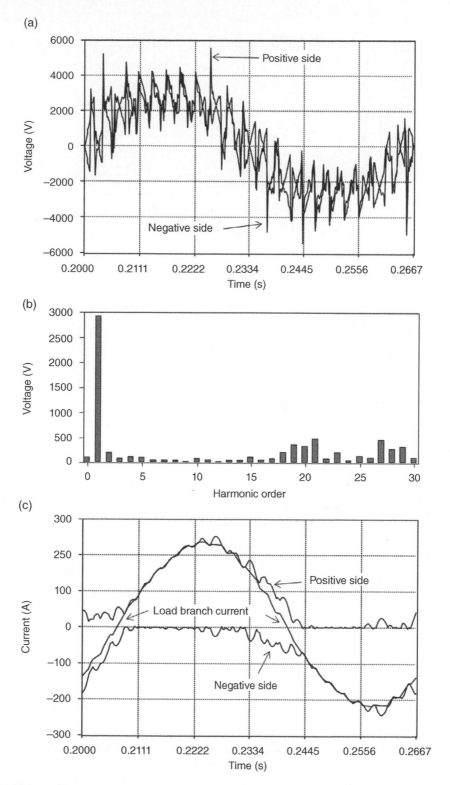

Figure 10.9 Cycloconverter variables. (a) Across-bridge voltages. (b) Fourier spectrum of the voltage. (c) Through-bridge and load currents.

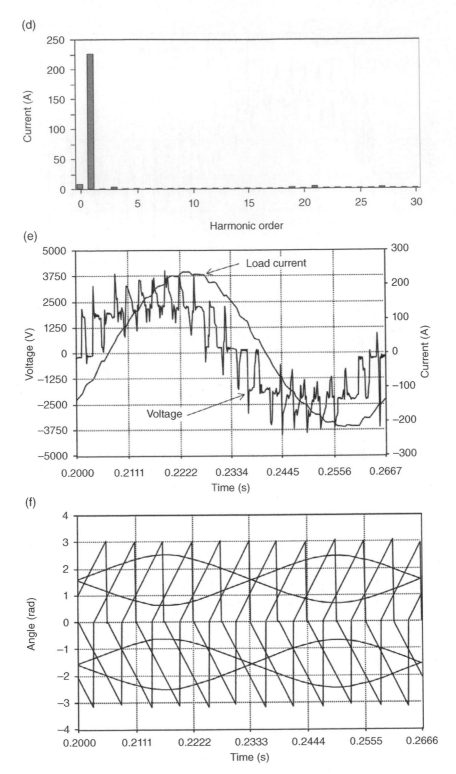

Figure 10.9 (*Continued*) (d) Fourier spectrum of the load current. (e) Voltage and current of load. (f) Ignition delay angle calculation basis.

Figure 10.9d shows the Fourier spectrum of the load current, which involves fewer harmonics, in other words, little distortion of the wave shape.

Figure 10.9e shows the load voltage and the current wave shapes. The instantaneous voltage in the power system, especially around the current zero time interval, is high. This corresponds to a very low power factor of load current in the power system. This is a typical shortcoming of cycloconverters.

Figure 10.9f shows the system (60 Hz) phase-to-phase voltage based sawtooth waves (three-phase) and cosine of the target voltage basis (15 Hz), by which the ignition delay angles (alphas) are calculated in TACS.

10.3 Cycloconverter-Driven Synchronous Machine

10.3.1 Application of Sudden Mechanical Load

Some rolling machines are driven by a synchronous motor, the power sources of which are currently cycloconverters. In this section, the cycloconverter-driven synchronous machine is discussed. In addition, comparison with an inverter-driven system will be shown.

The circuit layout shown in Figure 10.10 is applied. The following are to be noted:

- In Section 10.2, detail of one-phase cycloconverter is explained. Three of the same converter systems are applied for driving a three-phase synchronous machine, the ratings of which are 3.3 kV, 15 Hz, 1 MVA, 6P, and so on.

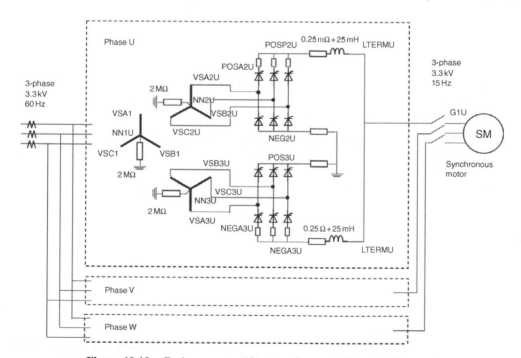

Figure 10.10 Cycloconverter-driven synchronous motor circuit layout.

- The transformer secondary side (converter valve side) is to be in the nonsolidly earthed condition in each phase. Therefore, two sets of high-ohmic resistor earthed star windings are applied for each phase, as shown in the figure. The primary side could be a common set of three-phase windings. In this case, for simplification, three sets of star-connected windings are applied for three-phase.
- Three-phase reference voltages are to be given to three-phase converter controlling (in TACS).
- For initialization, very fine tuning is required, especially between the reference voltage and the initial machine terminal voltage, regarding the amplitude and phase angle.

The first example is to apply a **sudden mechanical load** to the rotating motor in an almost noload condition. The initialization and transient calculation process applied is the following (as the most simplified one):

- Initially, the motor is disconnected from the cycloconverter.
- Automatic initialization is highly recommended for synchronous machines. The motor is rotating in very lightly loaded generator mode, that is, giving the terminal voltage with the relevant frequency, and a high-ohmic resistor is connected to the terminal of the machine for the purpose of easy and proper initialization.
- In the next step, the machine is connected to the cycloconverter source for motor operation.
- Afterward, a sudden mechanical load is applied, such as in rolling machines, by means of TACS.

  For details, see the data file.[6]
  Some calculation results are shown in Figure 10.11.

(a) As details are shown in another chapter, cycloconverter output voltage involves a certain amount of high-frequency components.
(b) Synchronous motor current, due to reactance components in the circuit, does not involve a significant amount of high frequency component. The current amplitude changes a lot depending on the load condition.
(c) Suddenly applied mechanical torque (TACS controlled) and calculated air gap torque are shown. Due to sudden application and ejection of the torque, significant swing of the air gap torque is produced.
(d) By the swing, the motor current (in total time range) changes a lot.
(e) The swing is observed in rotor position angle and velocity as well.
(f) A cycloconverter-driven system consumes a lot of reactive power. In many cases, compensation facilities (capacitor bank) are required.
(g) Positive and negative polarity converter bridges work well at turning over.
(h) At a highly loaded instant, the power factor seems to be high. Figure 10.11f clarifies this.
(i) At the load eject instant, the current drastically changes, especially in the phase angle. Turning over between the positive and negative bridges seems to be suitable from the current wave shape.

---

[6] ATPData10-3-1.dat: Cycloconverter-driven synchronous motor is initialized. Sudden mechanical torque is applied. The torque is suddenly dropped a little later.

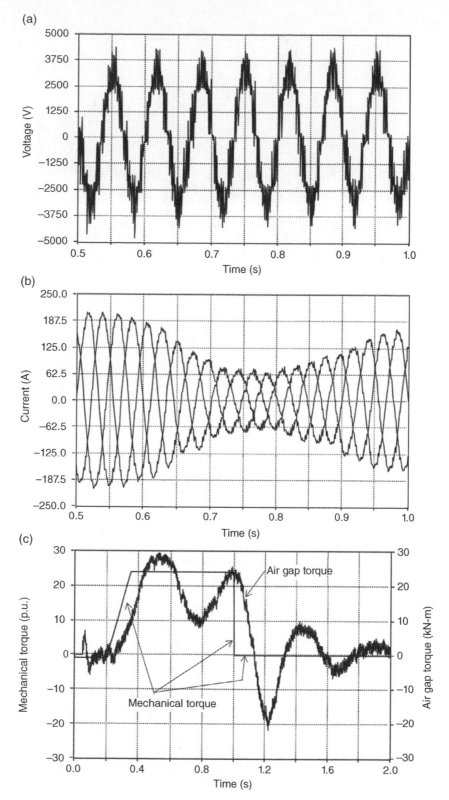

Figure 10.11 Cycloconverter-driven synchronous motor. (a) Cycloconverter output voltage. (b) Synchronous motor current. (c) Mechanical and air gap torques.

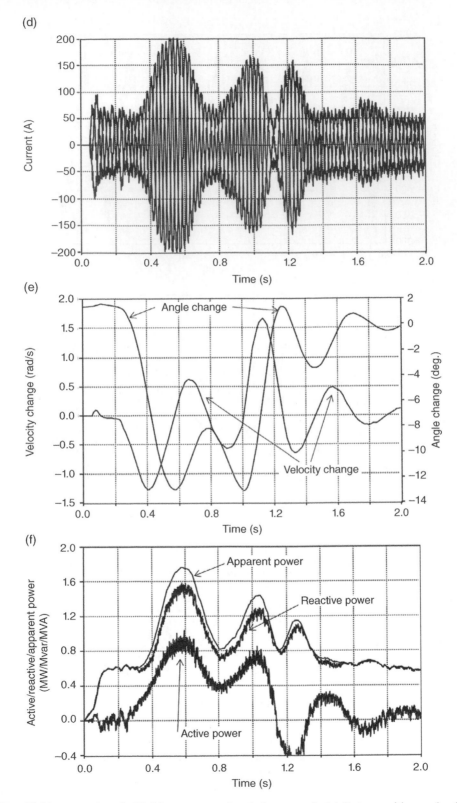

Figure 10.11 (*Continued*) (d) Motor current (total time range). (e) Rotor position and velocity. (f) Active/reactive/apparent powers.

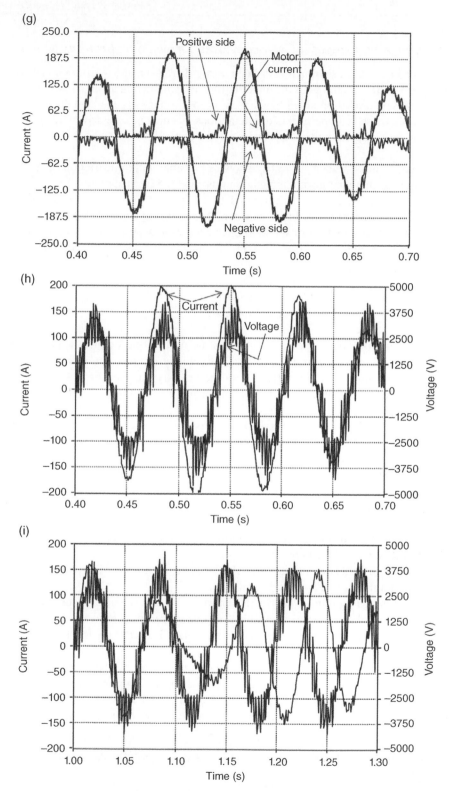

Figure 10.11 (*Continued*) (g) Cycloconverter output currents. (h) Current and voltage—highly loaded. (i) Current and voltage—load ejecting.

10.3.2   Quick Starting of a Cycloconverter-Driven Synchronous Motor

In the next example, quick starting of a cycloconverter-driven synchronous motor is discussed.

In Appendix 8.B, a synchronous machine starting as an induction machine is demonstrated, where, to represent a short-circuited field coil, very low voltage is generated during initialization. Now, in this case, a properly exited synchronous machine is to be driven by cycloconverter. Then, the following initialization process is to be applied:

- The machine's initial velocity is to be as low as possible within restriction of ATP-EMTP synchronous machine initialization menu. In this case, 0.5 Hz is applied.
- The motor-internally generated voltage is to be proportional to the velocity, that is, 0.5/15 = 0.033 times of the rated voltage where the rated frequency of the motor is 15 Hz.
- The applied voltage should correspond to the machine-induced voltage. So, cycloconverter output voltage should have linearly rising frequency and amplitude, corresponding to linearly rising velocity.

For details of the input data file, see data file.[7]
Some calculation results are shown in Figure 10.12.

(a) The command (reference) velocity, which is the base of the frequency and the voltage amplitude, and the resultant calculated motor velocity (represented in electrical angular velocity) are shown. Certain mitigation to the velocity change, for smooth mechanical response, is applied using s-block function in TACS. A little bit of instability in the motor velocity, especially in a higher velocity region, is observed.

(b) Three-phase reference voltage wave shapes. As shown in Section 10.2, cycloconverter output voltages correspond to these wave shapes. Linearly rising amplitude and frequency of waves are represented.

(c) Cycloconverter-created applied voltage and motor current of the one-phase is shown for total time range. The current amplitude is not constant.

(d) Detail of the initial section of part (c) of the figure is shown, where mainly reactive current flows.

(e) Detail of part (c) when there is higher torque outputting. The power factor seems to be higher, at least around the motor part.

(f) Detail of part (c), but after the start and while rotating by no-load. The power factor seems to be low and the current amplitude is also low.

(g) The motor air gap torque is shown in contrast with the velocity. Instability is clearly shown, most probably due to mechanical and/or electrical parameters. For a smoother response, further stability studies seem to be necessary. Powerful feedback control systems may be effective.

(h) Active power, reactive power, and apparent power measured at the power frequency inputting point are shown. At t = 0.9 s, when the motor torque is highest, the power factor not shown in the figure is still not so high, that is, around 0.3 p.u. For this reason, compensation facilities (capacitor bank) are installed by some cycloconverter systems.

---

[7] ATPData10-3-2.dat: The synchronous motor very quickly starts from the stalled condition by VVVF output of the cycloconverter.

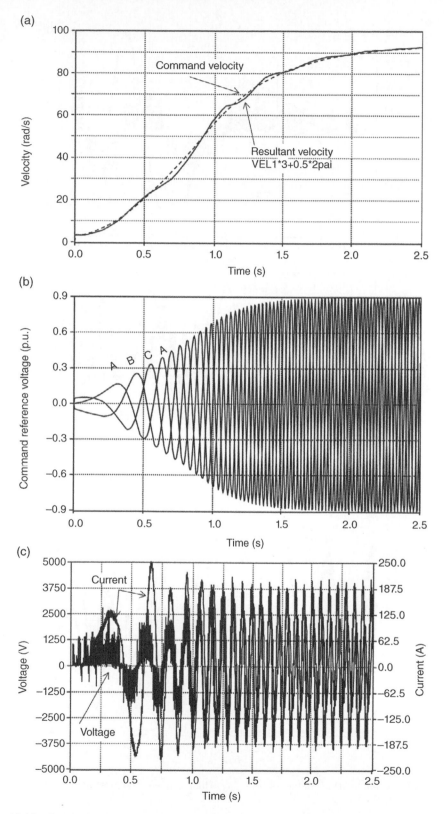

Figure 10.12 Quick starting of cycloconverter-driven synchronous motor. (a) Command (reference) and resultant velocities. (b) Reference voltage wave shapes. (c) Output voltage and motor current.

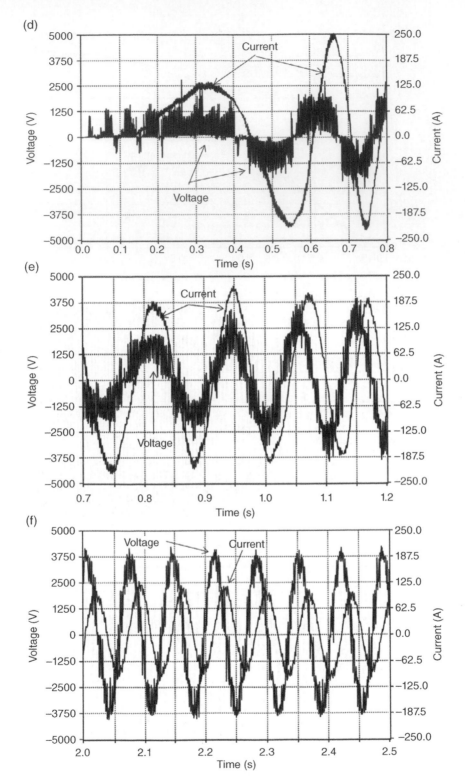

Figure 10.12 (*Continued*) (d) Initial part of (c). (e) High torque output time interval of (c). (f) Steady rotating time of (c).

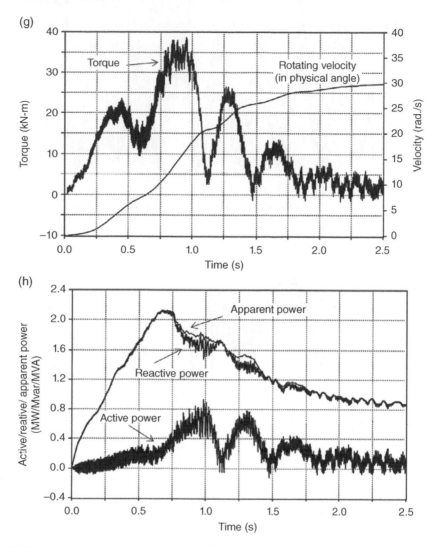

Figure 10.12 (*Continued*) (g) Torque and velocity. (h) Active/reactive/apparent power.

10.3.3 Comparison with the Inverter-Driven System

First, a "sudden mechanical load application" is introduced to the inverter-driven system. In the inverter-driven system, a circuit layout mostly equal to Figure 10.1c is applied. (For details, see data file.[8]) The reference voltage to control inverter output is identical to that in the cycloconverter.

Typical comparisons between the two are shown in Figure 10.13.

---

[8] ATPData10-3-3.dat: The same synchronous machine operating condition as ATPData10-3-1.dat case, but instead of a cycloconverter, a diode-bridge rectifier, and a PWM inverter are applied. The reference voltage to control the inverter is identical to that in ATPData10-3-1.dat case.

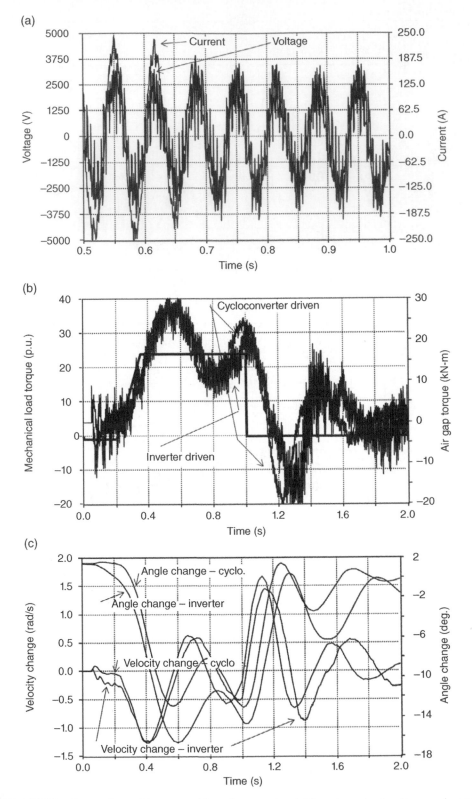

Figure 10.13 Comparison of inverter with cycloconverter. (a) Inverter output voltage and current. (b) Mechanical and air gap torques. (c) Rotor position angle and velocity changes.

(d)

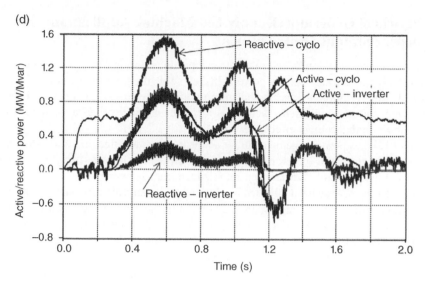

Figure 10.13   (*Continued*) (d) Active/reactive powers.

(a) Inverter output voltage and motor current for mainly heavy mechanical loading time intervals are shown. The power factor seems to be very high. For no-loading time intervals (not shown), on the other hand, the power factor at the motor input is very low.

(b) By sudden mechanical load torque, air gap torques by both cases are compared. Both are mostly identical, that is, the responses are almost equal from both source circuits.

(c) Little difference is observed in the rotor position angles from two sources. Though the reference voltage waves are identical to each other, small differences may be introduced between the two created voltage wave shapes. The cause might be the finite time step length of controlling.

(d) Great difference is shown in the power frequency source supplying inductive reactive power between two source circuits. From the inverter, power frequency side reactive power is negligibly small. As written in (a), for a no mechanical load time interval (especially after 1.5 s), the motor consumes some reactive power; nevertheless, the power frequency side reactive power is negligible. The inverter itself supplies reactive power from the capacitor in the circuit.

Note:

Power outputs by inverter are smoothed by s-block function in TACS. As the calculation principle is based on balanced three-phase variables, such mitigation is to be applied for variables with high frequency components.

As for a cycloconverter system with relatively low cost, low power frequency, necessity of capacitor bank, and limitation in output frequency and inverter system with higher cost, high power frequency, nonnecessity of compensation, and higher output frequency, not only qualitative but also quantitative comparison could be exercised by ATP-EMTP simulation, such as was shown previously.

10.4   Flywheel Generator: Doubly Fed Machine Application for Transient Stability Enhancement

As shown in Chapter 9, a doubly fed machine can produce, though only for a relatively short time interval, both active and reactive powers. In Chapter 8, the transient stability phenomenon is explained with relation to energy balance in the relevant power system. From these combined, applying a doubly fed machine as a flywheel generator to a power system, the transient stability enhancement effect is expected.

Figure 10.14 shows power system layout for analyzing such an effect in a single line diagram.

In the figure, one generator versus an infinite bus power system is identical to the one in Chapter 8. The doubly fed machine as a flywheel generator is identical to the one in Chapter 9. The current regulated inverter to energize the doubly fed machine rotor is identical to the one in Chapter 9.

In the figure, infinite DC voltage source is applied to the current regulated inverter power source. In actual cases, the DC source energy is to be supplied from the main power system. So the actual effect of the flywheel generator may be increased/decreased, depending on the velocity of the doubly fed machine. As in Chapter 9, it should be noted that the DC supplying system for the inverter is to be bidirectional, that is, a rectifying/regeneration system, as in a certain operating state the doubly fed machine rotor supplies energy toward the inverter side.

For controlling the flywheel generator to absorb/exhaust energy, usage of the information from the associated bus voltage is thought to be realistic. So in this case, frequency change of the bus voltage is picked up and applied to control the inverter, that is, the primary side generating power (both active and capacitive reactive) is set to be proportional to delta F. The detailed controlling algorism is shown in Figure 9.11. This output is applied as the reference current of the current-regulated inverter.

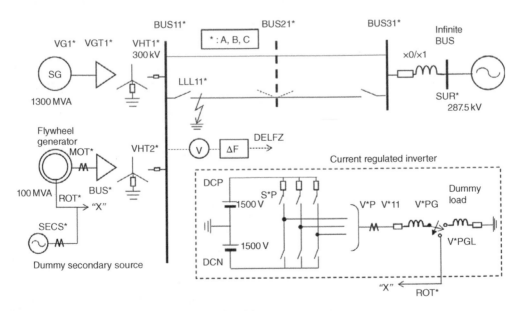

Figure 10.14 Single line diagram of a flywheel generator equipped generator versus infinite bus system for transient stability enhancement.

10.4.1 Initialization

Though each component in Figure 10.14 could be appropriately initialized in each respective way, combining plural components, a unified mode of initialization is to be applied according to the restriction of ATP-EMTP. In this case, step by step and/or trial and error procedures seem to be convenient.

Synchronous generator: In Chapter 8 where only synchronous generators are applied, the FIX SOURCES option is quite appropriately applied for each case initialization. However, in the case with universal machine(s), the option has not been successfully applied. Therefore, in this case, another mode that is also compatible with universal machines is to be applied to the synchronous generator initialization. As shown in Section 10.1, where a synchronous machine and cage rotor induction machine (universal machine) exist in a common system, inputting terminal voltage amplitude and phase angle for the synchronous machine, and slip value for the cage-rotor machine, produces appropriate initialization. So, in principle, let's try the same mode of initialization.

In the first step, the generator versus the infinite bus system (without flywheel generator) is automatically initialized applying the FIX SOURCES option for power flow calculation. (For details, see ATPData10-4-1.dat.) The initial generator terminal voltage amplitude and phase angle can be obtained.[9]

In the next step, excluding the FIX SOURCES option and applying the previously obtained voltage amplitude and phase angle to the generator terminal, the calculation is to be done. Very fine and precise tuning may be necessary, especially for the phase angle. For details, see ATPData10-4-2.dat. By the calculation, an identical result to the former automatic initialization case is obtained. Please compare both calculation results. Figure 10.15 shows some examples. A small amplitude of ripple in the air gap torque corresponds to the asymmetry of the transmission line (nontransposed).[10]

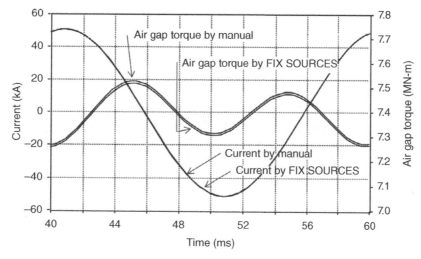

Figure 10.15 Precisely tuned generator's current and air gap torque compared to the automatic.

---

[9] ATPData10-4-1.dat: One synchronous generator versus infinite bus system automatically initialized by FIX SOURCES.

[10] ATPData10-4-2.dat: Manually initialized identically to the above condition.

In the third step, the flywheel generator is to be initialized. As shown in Chapter 9, the doubly fed machine is initialized best by introducing the primary side terminal voltage condition, slip value, and secondary side current condition. The primary side voltage condition is obtained from the bus voltage via the step-up transformer (Figure 10.14). The initial slip value is selected at 5% as, due to the primary stage of excess energy in the power system source part in transient stability phenomena, lower velocity seems to be appropriate to absorb the excess energy. By setting the flywheel initial condition as capacitor mode, and by very fine tuning of the secondary current condition, the machine is well initialized (See data file.[11]) Some results are shown in Figure 10.16. The main flux magnetizing current is supplied from the secondary side, so the amplitude of the secondary side current is higher. The secondary side current appears in the amount of 2.5 Hz (corresponding to 5% of slip). In this case, the secondary side is supplied by a fixed current source (2.5 Hz).[11]

The fourth step is to investigate the universal machine controlling reference current calculation algorithm shown in Figure 9.11. (For details, see the data file.[12]) A typical result is shown in Figure 10.17, which is a comparison of the actual (from outside source) and the calculated reference current that is to be applied to control the current regulated inverter. The inputting active/reactive power basis for the calculation is equal to the actual one. The difference between the two is negligible; though, due to mitigating the fluctuation in the calculated reference current, a small delay time is introduced. At any rate, the calculation algorithm seems to be quite agreeable for this purpose.

Then, activating the inverter, the flywheel generator's rotor is energized by the output current of the inverter. As shown in Chapter 9, the current regulated inverter output current is close to the input reference current. Therefore, applying the calculated reference current by

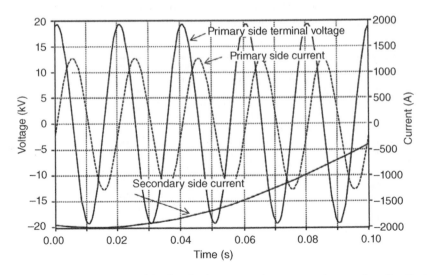

Figure 10.16   Initial stage of the flywheel generator (currents are in outgoing direction).

---

[11] ATPData10-4-3.dat: previous system plus flywheel generator, the rotor of which is energized by a fixed AC source outside.

[12] ATPData10-4-4.dat: Same as data file 11, but inverter control program is implemented and checked under the same condition. (Inverter nonactivated.)

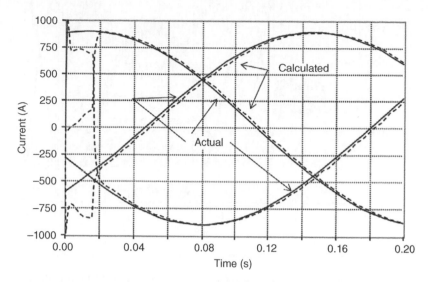

Figure 10.17 Check of the secondary current calculation algorithm by comparing the actual and the calculated reference current.

the previously shown calculation algorithm, the flywheel produces the target output power, the rotor being energized by the current of the inverter.

The basic flywheel generator output power is controlled to be the following (for details, see data file[13]):

- Initially low inductive reactive output power.
- At 0.2 s, the rotor input power is switched over from the fixed source to the inverter.
- At 0.4 s, the flywheel output is increased to −150 MW (active power absorbing, that is, motor mode).
- The calculation is continued up to 2 s.

Figure 10.18a shows the reference current to control the inverter and the machine rotor actual current (inverter output). Proper working of the inverter and proper driving control of the machine are expected. At around 1.2 s, the rotation of the current is reversed, that is, the velocity crosses the synchronous speed (52.36 rad/s; see Figure 10.18c).

Figure 10.18b shows machine-generating powers that are calculated from the terminal voltage and outgoing current. The active power is kept constant at −150 MW, though some fluctuations exist due to the inverter switching. The reactive power is negligibly small accordingly. Both powers are kept to the input condition. (See PP and QQ values in TACS of the data file.)

By absorbing active power at 0.44 s, the machine begins to accelerate from 49.74 rad/s (corresponding to +5% of slip) up to approximately 55 rad/s (approximately −5% of slip) at 2 s. Very accurately speaking, the acceleration rate should be inverse to the velocity due to the

---

[13] ATPData10-4-5.dat: Current regulated inverter activity is checked, energizing the flywheel rotor to create the same condition as ATPData10-4-3.dat.

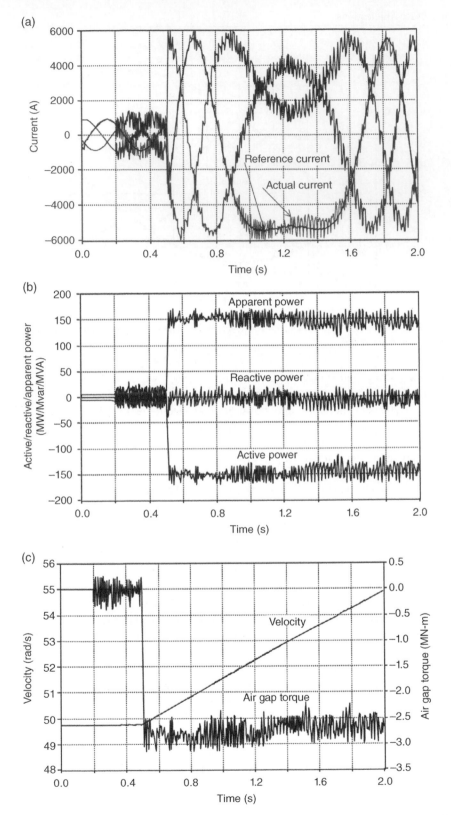

Figure 10.18 Activating inverter to drive flywheel generator. (a) Inverter activated: reference and actual flywheel currents. (b) Calculated active, reactive, and apparent powers. (c) Velocity and air gap torque.

constant active power value applied. The air gap torque has the same relationship. In Figure 10.18c, the air gap torque shows a gradual decrease in value by increase of velocity.

Some fluctuation also exists due to the inverter's switching.

Summing this up, the flywheel generator connected to the source side of the one generator versus the infinite bus system seems to be appropriately systematized and initialized. Introducing active and/or power value to the inverter control terminal, the flywheel outputs/absorbs power by changing the velocity.

The next subject is the flywheel's proper power control energized by the output current of the inverter. The synchronous generator's automatic voltage regulator/power system stabilizer (AVR/PSS), which is the most powerful means to enhance the transient stability, is generally controlled applying the generator's state, that is, the terminal voltage, output power, and so on. A flywheel generator may not be installed close to the synchronous generator. Moreover, plural synchronous generators may be covered by flywheel generator(s). Therefore, applying information directly regarding the synchronous generator may not be convenient. As mentioned at the beginning of this section, information regarding the voltage at the relevant bus close to the flywheel generator seems to be mostly applicable for this purpose.

As the most direct connection between the synchronous generator's disturbance and the bus voltage, the voltage frequency change seems to be applicable for the input of inverter control. Fortunately, we can use FREQUENCY METER in TACS of ATP-EMTP. So, at first the frequency meter should be checked during disturbance regarding transient stability (during 3LG and one circuit of the transmission line opening).

Applying FREQUENCY METER to the bus voltage where the flywheel generator is connected, and after some mitigating processes in the calculation, Figure 10.19's result is obtained. (For details of the calculation, see ATPData10-4-6.dat.[14])

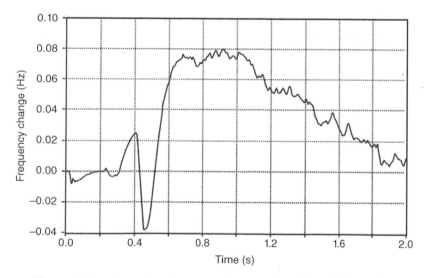

Figure 10.19   Bus voltage frequency change during 1LG–1CCT opening.

---

[14] ATPData10-4-6.dat: Total system in Figure 10.14 is checked where the flywheel is controlled by ΔF of the HV bus voltage under "3LG—1CCT opening" condition.

Due to the sudden change in the voltage during 0.3–0.4 s (3LG—Fault clearing—one circuit opening), the rapid frequency change in this time interval may be better excluded. However, rather steady frequency change output is obtained and suitable application to control the flywheel generator is expected.

10.4.2 Flywheel Activity in Transient Stability Enhancement

In the first trial, both active and reactive power outputs of the flywheel generator are set to be equal and proportional to the frequency change of the high voltage (HV) bus voltage, that is, when the frequency change is increased, the absorption of the active power and capacitive reactive power increase. Thus, the synchronous generator's acceleration is expected to be damped. The maximum power of the flywheel generator is set to be approximately 200 MVA, that is, 200% loading due to short time interval. Some results are shown in Figure 10.20.[15]

(a) Except for the violent transient interval (3LG and clearing), the flywheel rotor (secondary coil) current is well supplied by the inverter.
(b) According to the bus voltage frequency change (Figure 10.19) and along the vector control algorithm, the flywheel is driven to absorb power, enhancing the velocity, which is shown in Figure 10.20b.
(c) The figure shows the powers are well controlled. The same value of active and reactive powers is shown.
(d) The synchronous generator's d-axis angle (swing) during the transient is damped by the function of the flywheel. Due to the relatively low rate of the flywheel generator's power (as for active power, approximately 13% of the synchronous generator's), the damped rate is limited.
(e) The figure shows top (plus side) and bottom (minus side) valve currents in a certain phase of the inverter. For a certain time interval, only one side valve is on. This means the DC source voltage is critical and can never be lower.
(f) The HV bus voltage is lower during the swing. Therefore, according to Equation (8.3), the transmitting power is lower, resulting in excess source side energy. Absorbing higher capacitive reactive power to enhance the voltage, further damping of the swing may be expected.

10.4.3 Active/Reactive Power Effect

In the next case, mainly active power only is produced by the flywheel with an approximately equal value of the maximum apparent power to the previous case. In TACS of the data file, changing the coefficients of the active and reactive powers, the condition is easily introduced. (For details, see data file.[16])

Comparing to the previous case, some results are shown in Figure 10.21.

---

[15] ATPData10-4-7.dat: Active and reactive power output control (of the flywheel) case, where the apparent power output is approx. 200 MVA (200%) maximum.
[16] ATPData10-4-8.dat: Active power only control case, where the output is approx. 200 MW maximum.

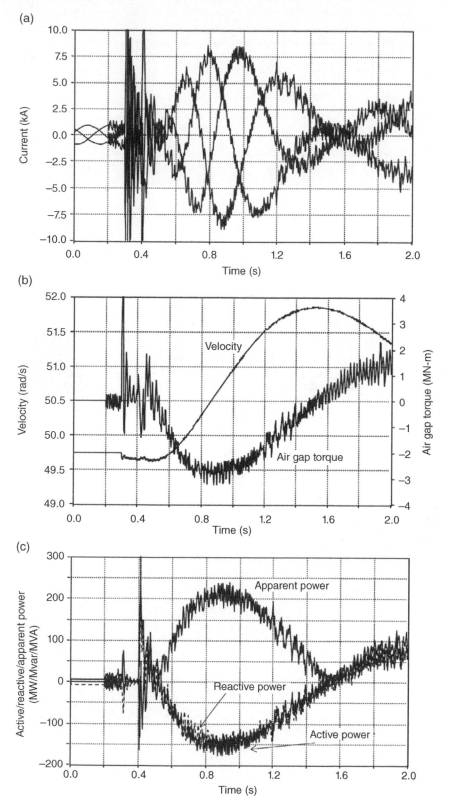

Figure 10.20 Flywheel generator activity in transient stability enhancement. (a) Rotor energizing current by inverter. (b) Flywheel velocity and air gap torque. (c) Flywheel output powers.

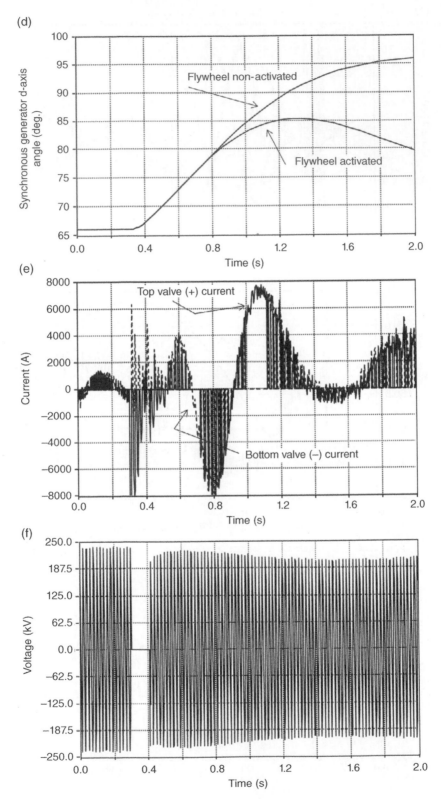

Figure 10.20 (*Continued*) (d) Synchronous generator d-axis angle. (e) Top and bottom valve switching over current. (f) HV bus voltage.

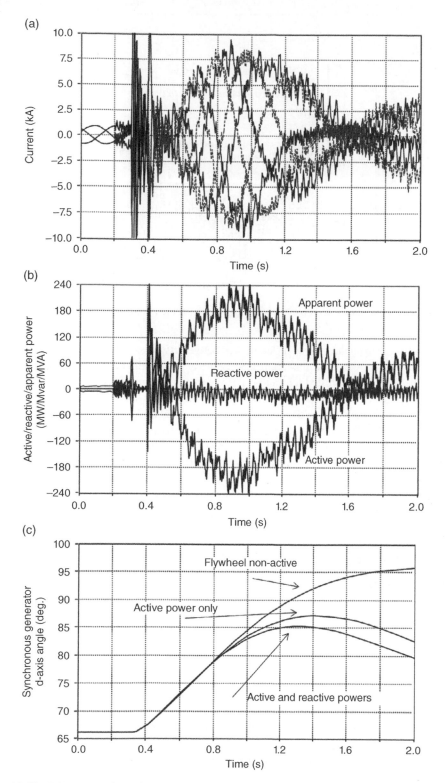

Figure 10.21 Effects by solo-active power and jointing with reactive power. (a) Rotor currents in both cases. (b) Active, reactive, and apparent powers. (c) Synchronous generator swing comparison.

(d)

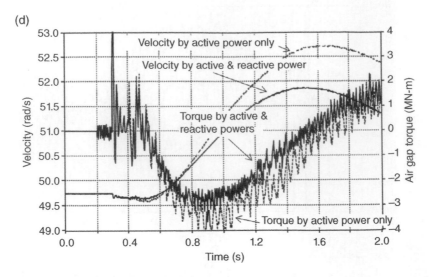

Figure 10.21 (*Continued*) (d) Velocity and air gap torque comparison.

(a) Rotor currents from the inverter are compared between two cases, where due to approximately equal apparent powers, the maximum crest values are approximately equal by two. The phase is, on the other hand, shifted relevantly. These can be explained by Figure 9.9.

(b) Active, reactive, and apparent powers of the flywheel are shown. Due to the almost zero reactive power, the value of the apparent power is equal to the active one, which is approximately equal to the previous case's apparent one (Figure 10.20c).

(c) Effects on the synchronous generator's swing (d-axis angle, which is the representative of the transient stability) are compared. In spite of approximately equal apparent powers, solo-active power control is not so effective compared to both active and reactive power control. A discussion will follow on this later.

(d) By solo-active power, velocity change and air gap torque are higher, though the effect is less.

10.4.4 Discussion

In Chapter 8, transient stability in the identical one generator versus infinite bus system is explained. Also, a significant effect of AVR/PSS to enhance stability is shown. In Figure 10.22a the effects of flywheel generator and AVR/PSS are compared under similar load flow conditions. The effect of AVR/PSS is apparently superior. With AVR/PSS, the synchronous machine's exciting is controlled, yielding enhancement of the transmission voltage. Therefore, with the increase of the transmission power according to Equation (8.3), the air gap torque rises as shown in Figure 10.22b. The maximum torque is 135% of the initial, and the difference from the initial torque, due to the constant mechanical input torque, acts to damp the swing. With the flywheel, the maximum torque is 118%, which is limited by the flywheel rating (overloading included). Moreover, the rise by AVR/PSS is far quicker. As a result, AVR/PSS is more effective in this case.

Nevertheless, depending on the system layout (the synchronous generator's and flywheel generator's ratings, fault conditions, etc.) various results are obtained [1].

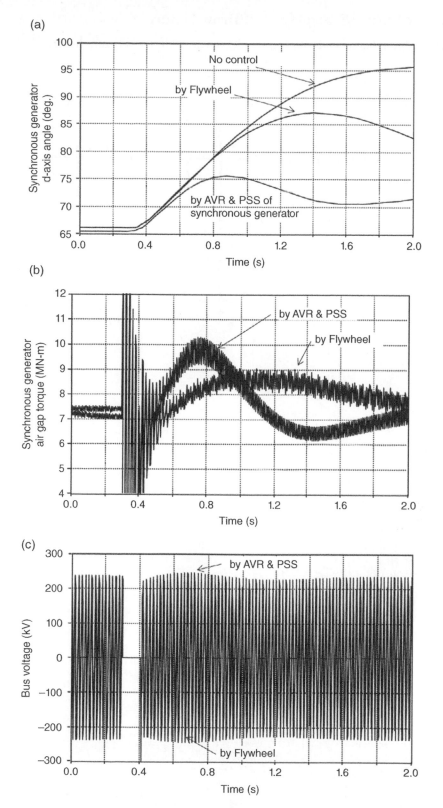

Figure 10.22 Comparison of flywheel with AVR/PSS for transient stability enhancement. (a) Synchronous generator's swing. (b) Synchronous generator's air gap torque. (c) HV bus voltage.

Appendix 10.A: Example of ATPDraw Picture

ATPDraw circuit diagram picture by Data10-1-1.acp[1] is shown in Figure A10.1.

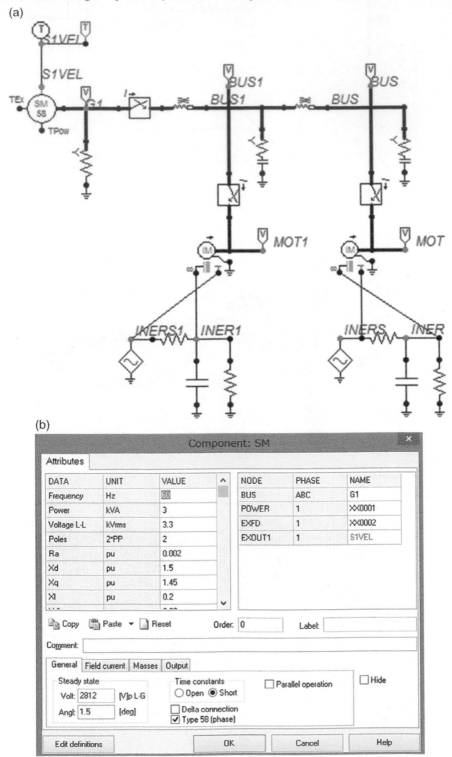

Figure A10.1   Basic small-size system. (a) Main circuit. (b) Synchronous machine parameters.

(c)

(d)

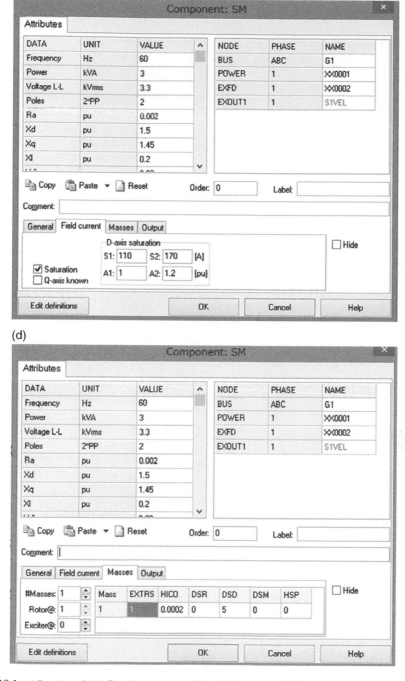

Figure A10.1 (*Continued*) (c) Synchronous machine field current. (d) Synchronous machine mass data.

(e)

(f)

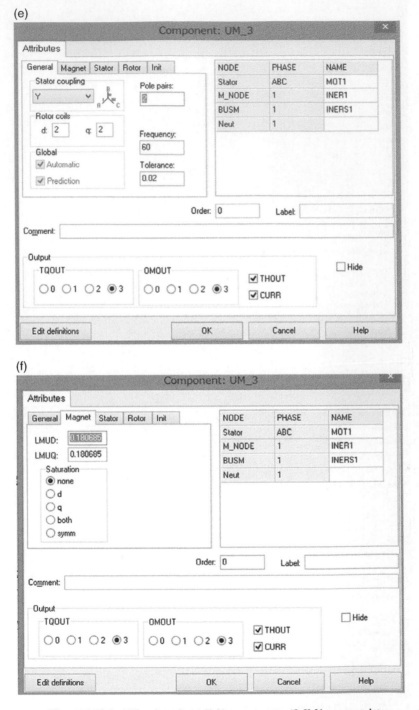

Figure A10.1 (*Continued*) (e) IM1 parameters. (f) IM1 magnet data.

(g)

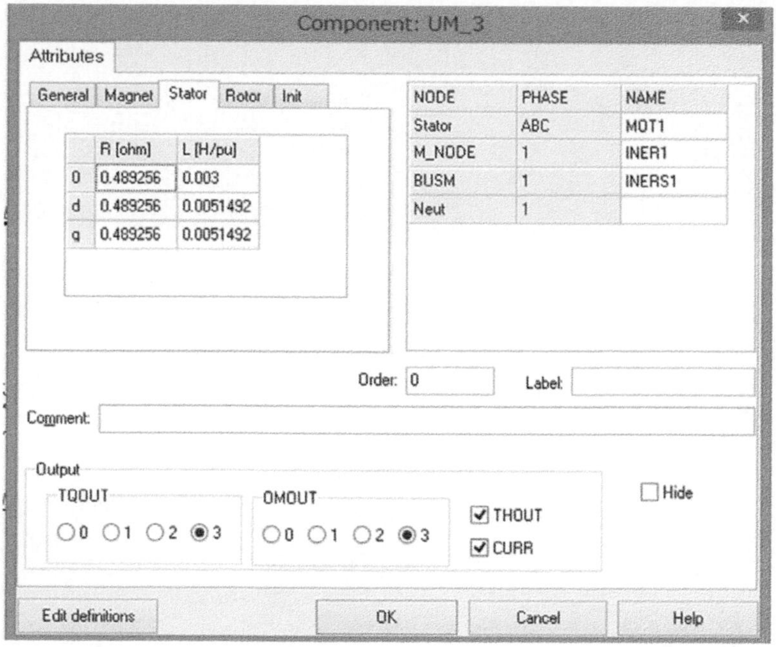

(h)

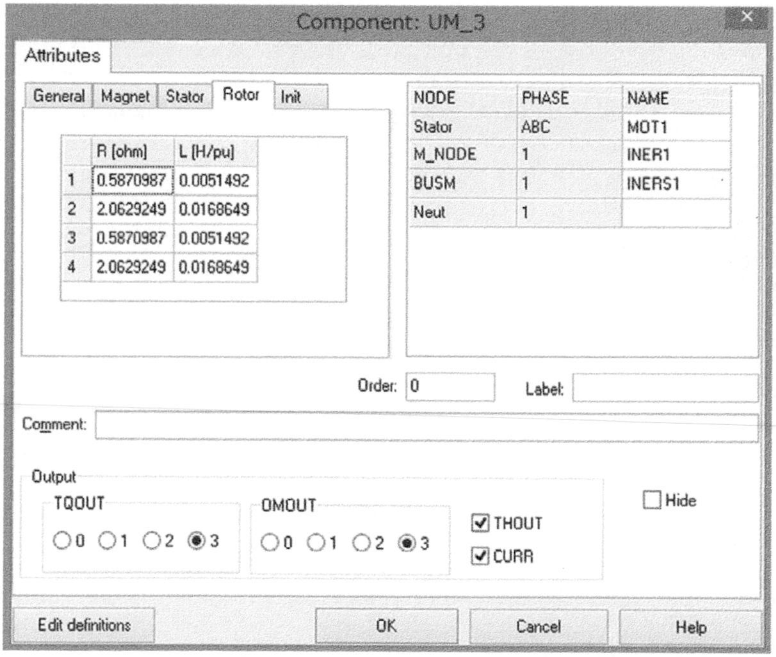

Figure A10.1 (*Continued*) (g) IM1 stator data. (h) IM1 rotor data.

(i)

(j)

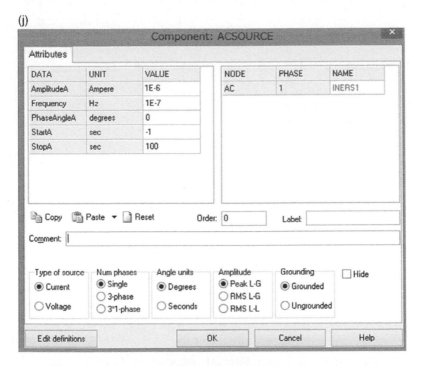

Figure A10.1   (*Continued*) (i) IM1 initial slip. (j) Current source.

(k)

(l)

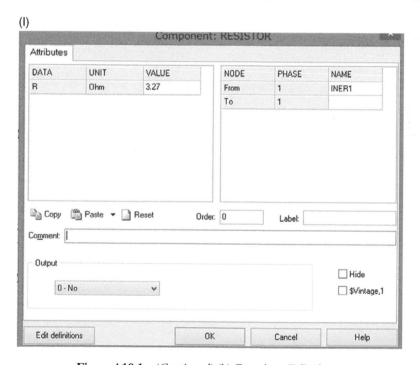

Figure A10.1 (*Continued*) (k) Capacitor. (l) Resistor.

Reference

[1] E. Haginomori (1998) EMTP simulation of transient stability enhancement phenomena by an inverter controlled flywheel generator, *Electrical Engineering in Japan*, **124** (3).

Index

Power System Transient Analysis: Theory and Practice using Simulation Programs (ATP-EMTP), First Edition.
Eiichi Haginomori, Tadashi Koshiduka, Junichi Arai, and Hisatochi Ikeda.
© 2016 John Wiley & Sons, Ltd. Published 2016 by John Wiley & Sons, Ltd.
Companion website: www.wiley.com/go/haginomori_Ikeda/power

Printed and bound by CPI Group (UK) Ltd, Croydon, CR0 4YY

07/02/2023

03189342-0001